Feature Paper Collection in Section 'Chromatographic Separations'

Feature Paper Collection in Section 'Chromatographic Separations'

Collection Editors

Victoria Samanidou
Evroula Hapeshi

MDPI

Basel • Beijing • Wuhan • Barcelona • Belgrade • Novi Sad • Cluj • Manchester

Collection Editors
Victoria Samanidou
Laboratory of Analytical Chemistry
Aristotle University of Thessaloniki
Thessaloniki
Greece

Evroula Hapeshi
Department of Health Sciences
University of Nicosia
Nicosia
Cyprus

Editorial Office
MDPI AG
Grosspeteranlage 5
4052 Basel, Switzerland

This is a reprint of the Topical Collection, published open access by the journal *Separations* (ISSN 2297-8739), freely accessible at: https://www.mdpi.com/journal/separations/topical_collections/0VI7U10QW5.

For citation purposes, cite each article independently as indicated on the article page online and as indicated below:

Lastname, A.A.; Lastname, B.B. Article Title. *Journal Name* **Year**, *Volume Number*, Page Range.

ISBN 978-3-7258-8507-7 (Hbk)
ISBN 978-3-7258-8508-4 (PDF)
https://doi.org/10.3390/books978-3-7258-8508-4

Contents

About the Editors

Victoria Samanidou

Victoria Samanidou is a Full Professor of Analytical Chemistry and Director of the Laboratory of Analytical Chemistry at the Department of Chemistry, Aristotle University of Thessaloniki, Greece. Her primary research interests focus on the development and validation of analytical methods using chromatographic instrumentation, specifically high-performance liquid chromatography and gas chromatography. She specializes in advanced sample preparation techniques, including modern solid-phase extraction and microextraction protocols applied to environmental, food, and biological matrices. She has coordinated and actively participated in numerous national and international research projects dedicated to the trace analysis of organic emerging contaminants, pharmaceuticals, and secondary metabolites using hyphenated mass spectrometry systems. Throughout her academic career, she has co-authored 239 original research articles in peer-reviewed journals, 76 review articles, 100 editorials/in view/opinions/commentaries/perspectives/forum articles, and 68 chapters in scientific books or tutorials. She is consistently recognized in the Stanford University database evaluating the top 2% of scientists worldwide based on citation impact. Furthermore, she serves on the editorial boards of several prominent scientific journals, contributes actively to the European Chemical Society (EuChemS) Division of Analytical Chemistry, and has served as President of the Association of Greek Chemists for the regional division of Central and Western Macedonia.

Evroula Hapeshi

Evroula Hapeshi has been an Assistant Professor in the Department of Health Sciences at the University of Nicosia since 2019. Her work centers on developing and applying advanced analytical methods based on UPLC–MS/MS and UPLC–QqTOF–MS for multi-residue analysis of organic compounds—particularly licit and illicit drugs and their metabolites, proteins, and peptides—in challenging matrices such as biological fluids and natural products. Recent projects include using drug and metabolite levels in wastewater as epidemiological biomarkers. She also designs and optimizes green-chemistry extraction protocols for compounds from natural products. She has actively participated in several national and European-funded research projects dedicated to the development of advanced hyphenated chromatographic methods, environmental monitoring, and the evaluation of advanced oxidation processes for wastewater treatment. Over the course of her career, she has published a substantial number of high-impact research articles in peer-reviewed scientific journals and presented her work at numerous international academic conferences. She is highly regarded for her contributions to environmental analytical chemistry and continuously serves as a peer reviewer and guest editor for leading scientific publications.

separations

MDPI

Article

Identification of Key Aroma-Active Compounds in Plum, Jujube, and Grape Extracts via GC×GC-O-TOF-MS, GC-MS, Aroma Recombination and Omission Tests

Ruiyang Chen [1,2,3], Qinghui Jia [1,2,3], Anzhen Fu [1,2,3], Fan Yang [1,2,3], Yixin Dai [1,2,3], Jingzhe Sun [1,2,3], Liya Liu [1,2,3], Ye Liu [1,2,3,*] and Shuang Bi [1,2,3,*]

1 Beijing Engineering and Technology Research Center of Food Additives, Beijing Technology and Business University (BTBU), Beijing 100048, China; rychen001015@163.com (R.C.); 2431031054@st.btbu.edu.cn (Q.J.); galaxyfaz@163.com (A.F.); 15733195507@163.com (F.Y.); daiyx6868@163.com (Y.D.); 18410163861@139.com (J.S.); 15011378004@163.com (L.L.)

2 Beijing Advanced Innovation Center for Food Nutrition and Human Health, Beijing Technology and Business University (BTBU), Beijing 100048, China

3 Flavor Science Laboratory, School of Food and Health, Beijing Technology and Business University (BTBU), Beijing 100048, China

* Correspondence: liuyecau@126.com (Y.L.); bishuang@btbu.edu.cn (S.B.)

Abstract

Fruity natural flavors are widely used in food, cosmetics, and tobacco products, and their flavor profiles directly affect the sensory quality of the final products. However, the volatile organic compound systems in these flavors are highly complex, and severe co-elution among volatile components makes effective component resolution a persistent analytical challenge. In this study, comprehensive two-dimensional gas chromatography–olfactometry–time-of-flight mass spectrometry (GC×GC-O-TOF-MS) was employed to separate and identify volatile odor compounds in three natural fruit flavor extracts: plum, jujube, and grape. A total of 113, 103, and 85 volatile odor compounds were identified from the plum, jujube, and grape extracts, respectively. By progressively adjusting the split ratio, aroma extract dilution analysis was performed, leading to the screening of 14, 11, and 32 aroma-active compounds from the three extracts, respectively. Compounds with high flavor dilution factors were subjected to quantitative analysis, and their odor activity values were calculated. Subsequently, omission and recombination experiments were conducted to confirm the key aroma-active compounds in each extract. The results showed that octanal was a key aroma compound shared by the plum and jujube extracts, while ethyl cinnamate was common to both the plum and grape extracts. Through a molecular sensory science approach and utilizing GC×GC-O-TOF-MS chromatographic separation technology, this study provides a reliable analytical platform for aroma characterization in complex samples. It also establishes a critical theoretical and data foundation for precisely identifying key aroma-active components and implementing targeted aroma modulation to enhance the natural flavor quality of fruits.

Keywords: fruity natural flavors; GC×GC-O-TOF-MS; molecular sensory science; separation; identification

1. Introduction

Flavors are substances that release specific aromatic components, perceived olfactorily to evoke pleasant sensations, and are classified as natural and synthetic flavors. Natural flavors are derived from animals and plants, such as flowers, leaves, fruits, resins, or animal secretions. Several plant-derived natural flavors are widely used in foods, beverages, confectionery, as well as in personal care products such as soaps, toothpaste, shampoos [1], and cosmetics (e.g., perfumes) to enhance appeal or mask undesirable tastes or odors. Moreover, plant-based natural flavors are applied as dietary supplements or added to functional foods to enhance nutritional value [2–5]. Natural flavor extracts, concentrated from plant-based aromatic materials and rich in complex volatile and nonvolatile components, serve as crucial, intensely flavored base materials for premium fragrances, cosmetics, and functional foods. The sensory quality of these extracts directly influences the grade, characteristics, and overall quality of downstream products.

The flavor profile serves as a key criterion for assessing the quality of flavor products. Although several studies have focused on the aroma profiles of various fruits, studies specifically addressing the flavor characteristics of fruit extracts remain limited. Plums, grapes, and jujubes are fruits characterized by sweet, sour, and fruity flavors. Hu et al. [6] identified 495 and 466 flavor compounds in the plum cultivars 'Fengtangli' and 'Siyueli,' respectively, and reported that 'Fengtangli' plums exhibited significantly higher relative contents of hydrocarbons, alcohols, ketones, and esters than 'Siyueli' plums. Additionally, 'Fengtangli' plums exhibited more pronounced sensory attributes, including sweet, citrus, herbal, floral, and fruity aromas. H. C. Liu et al. [7] examined plum flavor compounds across three maturity stages, identifying eight shared volatile odor compounds, including benzaldehyde, ethyl acetate, and ethyl propionate. Yue et al. [8] employed gas chromatography–mass spectrometry (GC-MS) to analyze flavor compounds in jujube at different developmental stages and found that aldehydes were the primary volatile aroma compounds, with (*E*)-2-hexenal being particularly abundant. Similarly, R. J. Liu [9] employed GC-MS combined with OAV (odor activity value) analysis and identified aldehydes, ketones, and esters as the main volatile aroma compounds in jujube fruit. The study revealed that the aroma profile was primarily characterized by fruity and fatty notes, with β-damascenone contributing to fruity characteristics and (*E*)-2-oct-en-1-al and nonanal significantly influencing the fatty aromas. Wu et al. [10] analyzed the volatile aroma compounds in three grape varieties with distinct aroma profiles, highlighting esters and aldehydes as key contributors to their fruity notes. Wang et al. [11] analyzed the odor profile of the widely cultivated grape variety 'Queen Nina', identifying benzeneacetaldehyde, 2,5-dimethyl-4-hydroxy-3(2H)-furanone, ethyl octanoate, ethyl valerate, and (*E*,*Z*)-2,4-dodecyl formate as the key volatile compounds associated with fruit ripening. Notably, the chemical composition of fruit extracts is significantly more complex than that of fresh fruit or juice. During concentration, extracts develop a complex matrix in which compounds that are insignificant in fresh fruit may become key contributors to aroma. Therefore, the direct application of analytical approaches and conclusions derived from fresh fruit studies to fruit extracts has inherent limitations.

However, most existing studies rely on conventional one-dimensional gas chromatography techniques, such as gas chromatography–olfactometry–mass spectrometry (GC-O-MS) or gas chromatography–ion mobility spectrometry (GC-IMS), which offer limited separation capacity for complex aroma systems and often result in insufficient data resolution. In contrast, comprehensive two-dimensional gas chromatography–olfactometry–time-of-flight mass spectrometry enables effective separation of complex mixtures by employing two chromatographic columns with different polarities, allowing analytes with similar physicochemical properties to be further resolved based on additional interactions [5]. Owing to its superior separation capabilities, the comprehensive two-dimensional gas chromatography–time-of-flight mass spectrometry (GC×GC-TOF-MS) has proven highly effective for flavor analysis in traditional Chinese medicines [12] and plant essential oils [13]. By separating analytes sequentially across two chromatographic columns with different polarities, GC×GC-TOF-MS achieves enhanced resolution of compounds with similar physical or chemical properties [14]. This study marks the first application of GC×GC-O-TOF-MS within a comprehensive molecular sensory science workflow to the identification of key aroma compounds in three types of natural fruit-flavored flavorings, enabling high-resolution analysis of complex volatile mixtures.

Although several volatile compounds are present in aroma systems, only a limited subset significantly influences characteristic aromas and determines the overall profile. These compounds are known as key aroma compounds. Studies on fruit-flavored fragrances have predominantly focused on characterizing volatile flavor compounds, frequently overlooking the identification of the aroma-active components. The screening of key aroma compounds necessitates the application of molecular sensory science approaches. Molecular sensory science integrates qualitative, quantitative, and descriptive molecular-level analyses, enabling the accurate reconstruction of flavors and the precise identification of compounds that contribute to aroma. Through continued methodological development, molecular sensory science has become one of the most advanced and systematic analytical frameworks for evaluating the flavor characteristics of fruits and fruit-derived products [15]. The application of molecular sensory science in identifying key aroma compounds provides significant practical benefits for enhancing the flavor profiles of natural fruit-flavored fragrances. Additionally, it establishes a robust scientific foundation to guide both product development and consumer-focused applications.

Therefore, the objectives of the present study were to: (1) screen aroma-active compounds in plum, jujube, and grape extracts using GC×GC-O-TOF-MS combined with FD factor analysis; (2) quantitatively analyze the identified aroma-active compounds in the three extracts using GC-MS; and (3) identify the key aroma-active compounds in plum, jujube, and grape extracts through omission and recombination experiments.

2. Materials and Methods

2.1. Materials and Reagents

Plum, jujube, and grape extracts were sourced from commercial suppliers. Chromatographic-grade methanol (purity $\geq$ 99%) used for the extraction and separation of flavor compounds was obtained from Beijing Mairuida Technology Co., Ltd. (Beijing, China). The internal standard quinoline (98%) was provided by Tokyo Chemical Industry. (Shanghai, China). Chromatographic-grade n-alkanes (C7-C40) for retention index determination were purchased from Sigma-Aldrich (St. Louis, MO, USA). The following reference standards were obtained from Shanghai Yuanye Biotechnology Co., Ltd. (Shanghai, China): octanal (99%), α-angelica lactone (97%), furfural (99%), benzaldehyde (99%), linalool (98%), 5-methyl-2-furaldehyde (98%), benzeneacetaldehyde (98%), 5-methyl-2-furanmethanol (95%), methyl phenylacetate (99%), guaiacol (98%), phenethyl alcohol (99%), pyrrole-2-carboxaldehyde (98%), ethyl cinnamate (98%), eugenol (98%), hexanal propylene glycol acetal (98%), isovaleraldehyde propylene glycol acetal (98%), ethyl hexanoate (99%), 1-octen-3-ol (98%), estragole (98%), geranylacetone (98%), 2-acetylpyrrole (99%), 4-ketoisophorone (98%), γ-nonalactone (98%), and γ-decalactone (99%).

2.2. Sensory Evaluation

The sensory evaluation was performed according to a previously described method [16] with slight modifications. The sensory panel was recruited from the Molecular Sensory Science Laboratory at Beijing Technology and Business University. The panel consisted of 12 members (6 male and 6 female) aged 22–30 years. Panelists underwent training in sensory evaluation techniques and descriptive quantitative analysis before the assessments. And all panelists signed an informed consent form, which was approved by the University

https://doi.org/10.3390/separations13060177

Scientific Research Ethics Committee of BTBU (2025-50). During the evaluation, panelists first described the odors of the natural flavor raw materials. Based on a subsequent panel discussion, the six most frequently cited sensory descriptors were selected, including "sweet," "fruity," "caramel," "floral," "licorice," and "sour." Sensory intensity was assessed using a 5-point scale, where "0" indicated an imperceptible aroma, and "5" indicated the highest intensity. The final results were calculated as the average scores for all sensory descriptors and presented as a radar chart.

2.3. Solid-Phase Microextraction (SPME)

A 1 g sample portion was placed into a 20 mL headspace vial, and 1 mL of saturated sodium chloride solution was added to facilitate the complete release of volatile aromas. Then, 1 μL of quinoline (1.093 μg/μL) was added as an internal standard. The vial was immediately sealed and equilibrated in a 60 °C water bath for 20 min. After equilibration, the SPME fiber was inserted into the headspace vial, positioned above the sample surface without touching the liquid, and allowed to adsorb volatile compounds at 60 °C for 40 min. The fiber was then retracted and inserted into the GC×GC-O-TOF-MS injection port, where desorption occurred at 230 °C for 5 min.

2.4. Comprehensive Two-Dimensional Gas Chromatography–Olfactometry–Time-of-Flight Mass Spectrometry (GC×GC-O-TOF-MS) Analysis

Volatile compounds in the three extracts were analyzed using an Agilent 8890A-5977B GC×GC-O-TOF-MS system (Agilent Technologies Co., Ltd., Shanghai, China) coupled with a Sniffer 9100 olfactometer (Gerstel Co., Ltd., Shanghai, China). The analysis was performed under the following conditions:

GC conditions: The system was equipped with two columns: a first-dimension DB-WAX polar column (60 m × 0.25 mm × 0.25 μm) and a second-dimension DB-17 ms mid-polarity column (1.85 m × 0.18 mm × 0.18 μm). An SSM1800 comprehensive (Xuejing Electronic Technology Co., Ltd., Shanghai, China) two-dimensional solid-state modulator controlled the heating and cooling stages between the columns, with the cold zone set at −51 °C and a modulation period of 5 s. The oven temperature program was as follows: an initial temperature of 40 °C held for 1 min, followed by a ramp of 2 °C/min to 230 °C, which was then held for 3 min. Samples were injected in splitless mode with helium as the carrier gas at a constant flow rate of 1 mL/min.

MS conditions: Mass spectrometry was performed using an electron impact (EI) ion source with 70 eV electron energy. The ion source temperature was 230 °C, the transfer line temperature was 280 °C, and the quadrupole was maintained at 150 °C. A solvent delay of 5 min was applied. Full-scan mode was used, with a mass range of m/z 35–450.

Olfactometry conditions: The transfer line to the olfactometry port was maintained at 200 °C. Humidified air was supplied to the sniffing port by bubbling nitrogen through a vial of ultrapure water. Olfactometry analysis was performed by three trained panelists, who recorded the perceived aroma attributes, intensities, and corresponding retention times.

2.5. Aroma Extract Dilution Analysis (AEDA)

Sample dilution was performed by adjusting the split ratio at the GC injector, with aromas diluted according to a geometric progression of 3^n. Three experienced sensory panelists assessed the aromas at the olfactometry port, recording the retention time, perceived aroma characteristics, and intensity for each compound. An aroma-active compound was considered present in the sample if it was detected by at least two evaluators. The FD factor of a compound was defined as the highest dilution at which its aroma could still be perceived. A higher FD factor indicates a greater contribution of the compound to the overall flavor profile of the sample.

2.6. *Qualitative Analysis*

Volatile aroma compounds were identified by integrating three complementary approaches: mass spectral matching, comparison of retention indices (RIs), and olfactometry (O) detection.

2.7. *Quantitative Analysis*

2.7.1. Semi-Quantitative Analysis

A semi-quantitative method using an internal standard was employed, with quinoline serving as the internal standard. The concentration of each target compound (Cx) was calculated using the following formula:

$$\mathrm{Cx} = \frac{\mathrm{Ax}}{\mathrm{Ai}} \times \mathrm{f} \times \mathrm{Ci}$$

where Cx represents the concentration of the target compound, Ci represents the concentration of the internal standard, Ax represents the peak area of the target compound, Ai represents the peak area of the internal standard, and f represents the correction factor.

2.7.2. Precise Quantitative Analysis

An internal standard calibration curve method was employed for the accurate quantification of aroma-active compounds with FD ≥ 3 in the three extracts. Mixed standard stock solutions of the aroma compounds were prepared at different concentration levels, with quinoline added as the internal standard for correction. Based on the selected ion monitoring (SIM) mode of the GC-MS system, standard curves were constructed by plotting the peak area ratio of the target compound to the internal standard (Y-axis) against the corresponding concentration ratio (X-axis) [17].

2.8. *Calculation of Odor Activity Values (OAVs)*

The OAV quantifies the contribution of an aroma-active compound to the overall flavor [18]. The OAV is defined as the ratio of the concentration of the compound in the sample to its odor threshold. Compounds with an OAV ≥ 1 are considered potential contributors to the aroma profile, and the relative contribution is proportional to the OAV [19]. The OAV was calculated using the following formula:

$$\mathrm{OAV} = \frac{\mathrm{C}}{\mathrm{OT}}$$

where C represents the concentration of the aroma compound in the sample, and OT represents the odor threshold of the compound.

2.9. *Aroma Recombination and Omission Experiments*

Recombination models were prepared by mixing aroma-active compounds with OAV ≥ 1 into a 1,2-propanediol matrix at their accurately determined concentrations. Due to its high solubility and chemical stability, 1,2-propanediol was selected, allowing the aroma-active compounds to mix uniformly and form a stable system. Sensory evaluations were performed on the three extracts and their corresponding recombination models.

For the omission experiments, mixture models were prepared by systematically omitting one aroma-active compound at a time from the complete set of identified compounds in each extract. Differences between the omission models and the complete recombination model were evaluated using triangle tests in accordance with the Chinese National Standard GB/T 12311-2012 [20]. Specifically, if 8, 9, or 10 out of 12 sensory panelists correctly identified a difference between the samples, the result was considered significant ($p \leq 0.05$),

https://doi.org/10.3390/separations13060177

highly significant ($p \leq 0.01$), or extremely significant ($p \leq 0.001$), respectively. If fewer than 8 panelists detected a difference, the result was considered non-significant.

2.10. Data Analysis

All experiments were performed in triplicate, and results are expressed as mean ± standard deviation. Statistical analyses were performed using Microsoft Excel 2023, and graphs were generated using Origin 2023.

3. Results and Discussion

3.1. Sensory Evaluation Analysis

The sensory profiles of the plum, jujube, and grape extracts were evaluated, and the results are shown in Figure 1. The plum extract exhibited pronounced sweet and sour characteristics. The jujube extract was characterized primarily by sweet and fruity notes. The grape extract also showed significant sweet and fruity notes. Among the three extracts, the grape extract exhibited the most pronounced fruity characteristics.

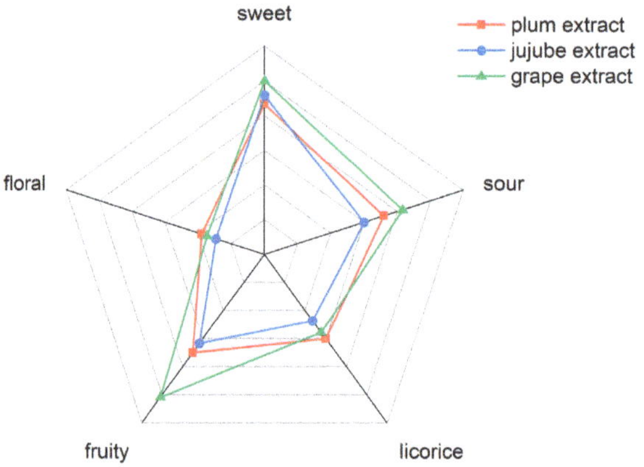

Figure 1. Sensory evaluation scores of three fruity natural flavor extracts.

3.2. Analysis of Volatile Aroma Compounds in the Three Extracts

A total of 188 volatile odor compounds were identified in the plum, jujube, and grape extracts using GC×GC-O-TOF-MS (Table A1 and Figure 2), including 44 esters, 35 aldehydes, 29 alcohols, 29 ketones, 22 heterocyclic compounds, 7 acids, 6 phenols, 5 alkenes, and 11 other compounds. Among these, 113 volatile odor compounds were identified in the plum extract, with alcohols, aldehydes, and esters being the dominant classes. Although alcohols and aldehydes were present in similar numbers, aldehydes exhibited significantly higher concentrations than alcohols. Notably, 5-methyl-2-furfural exhibited the highest concentration (111.099 μg/g), followed by 5-hydroxymethylfurfural (40.559 μg/g). A total of 103 volatile odor compounds were identified in the jujube extract, predominantly comprising aldehydes, esters, and heterocyclic compounds. The grape extract contained 85 volatile aroma compounds, mainly esters, aldehydes, and ketones. Esters were the most abundant class in the grape extract, with ethyl palmitate exhibiting the highest concentration (34.961 μg/g), which contributed to its characteristic fruity aroma.

https://doi.org/10.3390/separations13060177

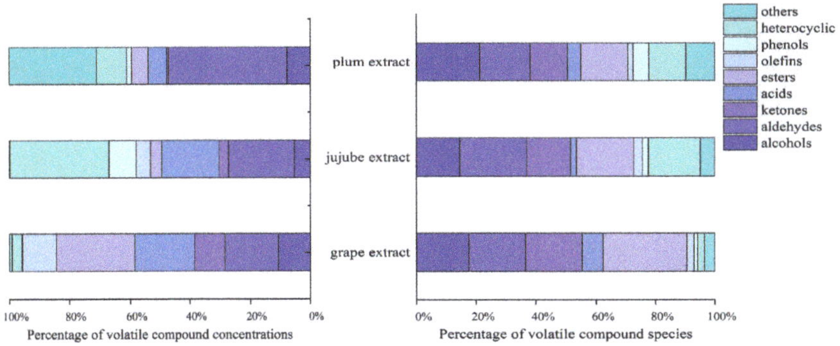

Figure 2. Comparison of volatile compound concentrations (**left**) and species (**right**) in three extracts.

3.3. Analysis of Aroma-Active Compounds in the Three Extracts

Aroma-active compounds directly influence the characteristic flavor profile of food products [21]. AEDA is widely used to evaluate the contribution of individual aroma compounds to the overall aroma, with higher FD values indicating a greater contribution of a given compound to the aroma intensity of the sample [22]. In total, 51 aroma-active compounds were identified across the three extracts (Table A2), comprising 15 aldehydes, 11 alcohols, 11 esters, 5 ketones, 4 acids, 2 phenols, 2 heterocyclic compounds, and 1 compound classified as other. Among the extracts, 14, 11, and 32 aroma-active compounds were identified in the plum, jujube, and grape extracts, respectively, indicating that the grape extract exhibited a more diverse profile.

3.3.1. Aldehyde

In the plum extract, two aldehydes with relatively high FD factors were identified: pyrrole-2-carboxaldehyde (FD = 27) and benzeneacetaldehyde (FD = 9). The compound pyrrole-2-carboxaldehyde is associated with a musty odor, whereas benzeneacetaldehyde contributes a sweet aroma. In the jujube extract, two aldehydes with high FD factors were identified: benzeneacetaldehyde (FD = 9) and octanal (FD = 3). Octanal is characterized by a waxy odor. Jia et al. [23] reported that octanal is a key aroma-active compound in dried jujube. In the grape extract, isovaleraldehyde propylene glycol acetal exhibited a significantly high FD factor of 729 and was identified as a major contributor to the fruity note.

3.3.2. Alcohols

Three alcohols with relatively high FD factors (FD = 9) were identified in the plum extract: linalool, 5-methyl-2-furanmethanol, and phenylethyl alcohol. Both linalool and phenylethyl alcohol are associated with rosy and floral notes, whereas 5-methyl-2-furanmethanol contributes a sweet aroma to the plum extract. Linalool has previously been confirmed as an aroma-active compound in plum [24]. In the jujube extract, alcohols with higher FD factors also include 5-methyl-2-furanmethanol. In the grape extract, 1-octen-3-ol exhibited a high FD factor (FD = 9) and is characterized by mushroom-like and earthy notes. It is generally formed via the lipoxygenase-mediated oxidation of linoleic acid [25].

3.3.3. Esters

The ester with the highest FD factor identified in both the plum and grape extracts was ethyl cinnamate (FD = 729). Ethyl cinnamate, which is characterized by a sweet aroma, was the dominant aroma-active compound in both extracts, consistent with pre-

https://doi.org/10.3390/separations13060177

vious findings [26]. In the jujube extract, the ester compounds with high FD factors were γ-decalactone (FD = 27), γ-nonanolactone (FD = 9) and coconut aldehyde (FD = 9). γ-Decalactone imparts a pronounced fruity aroma and was the major contributor to the overall aroma of the jujube extract, whereas γ-nonanolactone and coconut aldehyde contributed sweet notes to the aroma profile.

3.3.4. Heterocyclic Compounds

No heterocyclic compounds with high FD factors were detected in the plum and jujube extracts. In contrast, a high-FD heterocyclic compound, 2-acetylpyrrole (FD = 243), was identified in the grape extract and is characterized by a musty odor.

3.3.5. Phenols

Two phenolic compounds, guaiacol and eugenol, were identified in the plum extract, each exhibiting a high FD factor of 243. Guaiacol is characterized by a woody aroma, while eugenol has a sweet note. Guaiacol has been previously identified in plum [27], and eugenol has been confirmed as the primary aroma-active compound in plum, exhibiting an OAV greater than 1 [24]. In contrast, no phenolic compounds with high FD factors were detected in the jujube or grape extracts.

3.3.6. Other Compounds

No aroma-active compounds with high FD factors in the "Other" category were detected in the plum or jujube extracts. In contrast, estragole exhibited a high FD factor (FD = 729). Estragole, characterized by a sweet aroma, was a major contributor to the overall aroma profile of the grape extract, and its presence in grape skin has been previously reported [28].

3.4. Quantification of Aroma-Active Compounds and Calculation of OAVs

Aroma-active compounds with FD ≥ 3 in the three extracts were accurately quantified using the internal standard calibration curve method, and their OAVs were calculated (Table A3). The OAV represents the contribution of an aroma-active compound to the overall flavor rather than its concentration alone [29]. Compounds with OAV ≥ 1 are considered significant contributors to the aroma of the three extracts.

In the plum extract, 14 aroma-active compounds had OAV ≥ 1. Among them, methyl phenylacetate exhibited the highest OAV (502,747), followed by 5-methyl-2-furanmethanol (16,749), both of which primarily contributed sweet notes. Methyl phenylacetate has been previously detected in fruit pomace at concentrations exceeding its odor threshold [30]. In the jujube extract, six aroma-active compounds showed OAV ≥ 1, with 5-methyl-2-furanmethanol exhibiting the highest OAV (39,958). The compound 5-methyl-2-furanmethanol, which is formed during heating, was a major contributor to the sweet aroma of the jujube extract. In the grape extract, eight aroma-active compounds had OAV ≥ 1. Isovaleraldehyde propylene glycol acetal had the highest OAV (57,564,641), followed by 4-methyl-2-pentyl-1,3-dioxolane (355,960), both of which enhance the characteristic fruitiness of the grape extract.

3.5. Verification by Aroma Recombination and Omission Experiments

3.5.1. Aroma Recombination Experiment

Aroma recombination experiments are commonly used to validate the accuracy of identified key aroma-active compounds [31]. Recombination models were prepared by mixing compounds with OAV ≥ 1. The concentrations of aroma-recomposed compounds are presented in Table 1. As shown in Figure 3, the aroma profiles of the recombination models closely resembled those of the original extracts, indicating that the recombination models successfully simulated the overall flavor characteristics of the extracts.

Table 1. Concentration of aroma-recomposed compounds.

No.	Compounds [a]	Concentration (μg/g)
Plum extract		
1	Octanal	17.917 ± 1.327
2	Angelica lactone	454.138 ± 36.520
3	Furfural	297,076.482 ± 24,444.384
4	Benzaldehyde	1815.849 ± 218.245
5	Linalool	672.494 ± 73.571
6	5-Methyl-2-furanmethanol	147.397 ± 18.844
7	Benzeneacetaldehyde	521.623 ± 63.830
8	5-Methyl-2-furfural	47,857.922 ± 5856.231
9	Methyl phenylacetate	221.712 ± 22.081
10	Guaiacol	542.220 ± 72.182
11	Phenethyl alcohol	986.834 ± 59.631
12	Pyrrole-2-carboxaldehyde	2468.591 ± 225.528
13	Ethyl cinnamate	55.293 ± 8.569
14	Eugenol	28.572 ± 4.464
Jujube extract		
1	Octanal	90.921 ± 6.82
2	Benzeneacetaldehyde	440.280 ± 46.822
3	γ-Decalactone	65.479 ± 8.303
4	γ-Nonanolactone	289.821 ± 39.777
5	Ketoisophorone	146.435 ± 16.183
6	5-Methyl-2-furanmethanol	351.631 ± 30.605
Grape extract		
1	4-Methyl-2-pentyl-1,3-dioxolane	3210.760 ± 259.337
2	Isovaleraldehyde propylene glycol acetal	9958.683 ± 700.445
3	Ethyl caproate	2225.610 ± 181.657
4	Octanal	72.586 ± 10.384
5	1-Octen-3-ol	1882.864 ± 238.500
6	Estragole	337.489 ± 31.578
7	Geranylacetone	2064.986 ± 169.471
8	Ethyl cinnamate	728.362 ± 160.313
9	Artemisia ketone	337.489 ± 31.578

[a]: aroma-active compounds with OAV greater than 1.

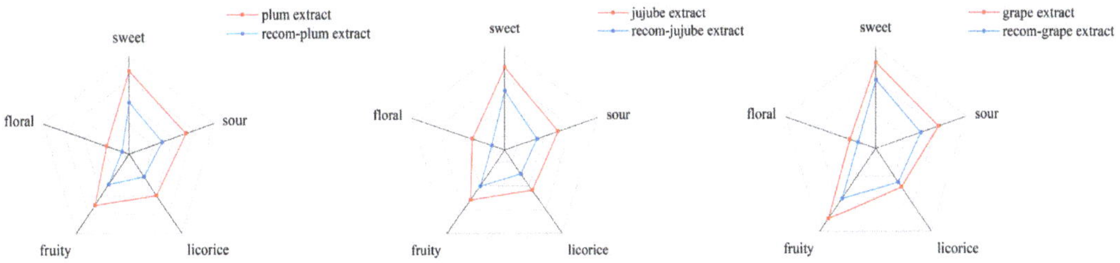

Figure 3. Sensory evaluation scores of three fruity natural flavor extracts and their aroma recombination model.

3.5.2. Aroma Omission Experiment

Aroma omission experiments were conducted to evaluate the contribution of major aroma-active compounds to the overall odor profile [19]. The results of the omission experiments are presented in Table 2.

Table 2. Triangle test results by omission experiments.

No.	Compounds [a]	n [b]	Significant [c]
Plum extract			
1	Octanal	8	*
2	Angelica lactone	8	*
3	Furfural	4	-
4	Benzaldehyde	3	-
5	Linalool	8	*
6	5-Methyl-2-furfural	8	*
7	Benzeneacetaldehyde	2	-
8	5-Methyl-2-furanmethanol	3	-
9	Methyl phenylacetate	9	**
10	Guaiacol	4	-
11	Phenethyl alcohol	1	-
12	Pyrrole-2-carboxaldehyde	8	*
13	Ethyl cinnamate	8	*
14	Eugenol	4	-
Jujube extract			
1	Octanal	8	*
2	Benzeneacetaldehyde	9	**
3	γ-Decalactone	8	*
4	γ-Nonanolactone	8	*
5	Ketoisophorone	8	*
6	5-Methyl-2-furanmethanol	10	***
Grape extract			
1	4-Methyl-2-pentyl-1,3-dioxolane	8	*
2	Isovaleraldehyde propylene glycol acetal	8	*
3	Ethyl caproate	2	-
4	Octanal	2	-
5	1-Octen-3-ol	9	**
6	Estragole	9	**
7	Geranylacetone	8	*
8	Ethyl cinnamate	8	*

[a]: aroma-active compounds with OAV greater than 1; [b]: number of correct judgements from 12 assessors by triangle test; [c]: significance: *, significant ($p \leq 0.05$); **, highly significant ($p \leq 0.01$); ***, very highly significant ($p \leq 0.001$); -, nonsignificant.

In the plum extract, methyl phenylacetate ($p \leq 0.01$), octanal ($p \leq 0.05$), angelica lactone ($p \leq 0.05$), linalool ($p \leq 0.05$), 5-methyl-2-furfural ($p \leq 0.05$), pyrrole-2-carboxaldehyde ($p \leq 0.05$), and ethyl cinnamate ($p \leq 0.05$) were identified as key aroma compounds. Among these, methyl phenylacetate, angelica lactone, linalool, and 5-methyl-2-furfural primarily contributed to the sweet notes of the plum extract.

In the jujube extract, 5-methyl-2-furanmethanol ($p \leq 0.001$), benzeneacetaldehyde ($p \leq 0.01$), octanal ($p \leq 0.01$), γ-decalactone ($p \leq 0.05$), γ-nonanolactone ($p \leq 0.05$), and ketoisophorone ($p \leq 0.05$) were identified as key aroma compounds. Among these, 5-methyl-2-furanmethanol, benzeneacetaldehyde, γ-nonanolactone, and ketoisophorone contributed to the sweet notes of the jujube extract. Notably, octanal was a key compound shared by both plum and jujube extracts. The octanal has a C_8 linear aldehyde structure; the carbonyl oxygen forms a hydrogen bond with the receptor, while the hydrophobic chain enters the receptor's hydrophobic pocket, activating specific olfactory receptors and producing a waxy sensation. Its FD factor is 3 in both plum and jujube extracts, with OAVs of 31 and 155, respectively. Although it does not dominate the aroma, it is an indispensable aromatic component in both. Meanwhile, octanal is a food fragrance used in the preparation of fruit flavors.

https://doi.org/10.3390/separations13060177

In the grape extract, 1-octen-3-ol ($p \leq 0.01$), estragole ($p \leq 0.01$), 4-methyl-2-pentyl-1,3-dioxolane ($p \leq 0.05$), isovaleraldehyde propylene glycol acetal ($p \leq 0.05$), geranylacetone ($p \leq 0.05$), and ethyl cinnamate ($p \leq 0.05$) were identified as key aroma-active compounds. Among these, estragole, 4-methyl-2-pentyl-1,3-dioxolane, isovaleraldehyde propylene glycol acetal, and ethyl cinnamate contributed primarily to the sweet notes. Geranylacetone was responsible for the sweet–fruity character of the grape extract. Ethyl cinnamate was also identified as a key aroma compound common to both the plum and grape extracts. In plant metabolism, phenylalanine serves as the primary volatile precursor for ethyl cinnamate [32]. The molecule of ethyl cinnamate consists of a benzene ring, a double bond, and an ester group; the absence of a p-methoxy group is key to its intense fruity and sweet aroma. The conjugated structure of ethyl cinnamate confers moderate lipophilicity and volatility, enabling it to easily penetrate the olfactory epithelium; once it reaches the receptors, π-π stacking, hydrogen bonding, and hydrophobic interactions work in concert to produce a strong and persistent sweet aroma even at extremely low concentrations. Its FD factor is 729 in both plum and grape extracts, with OAVs of 38 and 502, respectively, making it a key contributor to the sweet aroma. It is worth noting that ethyl cinnamate is classified as a nature-identical flavor ingredient. By adding ethyl cinnamate as a sweet aroma enhancer, it is possible to develop low-sugar foods and eco-friendly products free of artificial flavors and sweeteners. Additionally, it can be incorporated into cosmetics and personal care products to provide a fragrant experience. When added to tobacco, ethyl cinnamate helps neutralize the bitterness and pungency of the tobacco.

To verify the accuracy of the quantitative results, we selected two key aroma-active compounds with medium to high concentrations from each extract and conducted spiking recovery experiments. The recovery rates for the high-concentration compound phenethyl aldehyde and the medium-concentration compound thearubigone in the jujube extract both reached 89%, the recovery rates for the high-concentration compound furfural and the medium-concentration compound thearubigone in the plum extract reached 104% and 115%, respectively, and the recovery rates for isovaleraldehyde propylene glycol acetal and hexanal propylene glycol acetal in the grape extract were 105% and 119%, respectively, indicating that this method is accurate and reliable. The recovery rates of the selected representative compounds all fell within the acceptable range of 80–120%, indicating that this method is accurate and reliable.

Based on the identified key aroma compounds, these can also be used to develop natural fruit flavor formulations for plums, grapes, and jujubes. The process involves selecting aroma compounds with a high OAV as the primary components and using those with a low OAV to supplement and adjust the aroma profile. The ratios are evaluated and fine-tuned using scent strips to ultimately formulate the fruit flavors. Once the base formulation for the fruit flavor has been established, specialized secondary formulations must be developed for different product forms (food and beverages, cosmetics, perfumes, and tobacco products) and application scenarios.

In the food industry, high-OAV compounds can be used to replace natural extracts, enabling low-cost, standardized production. At the same time, by adjusting the ratios in flavor formulations, it is possible to develop fruit-flavored products with a variety of distinct flavors. In the cosmetics industry, these compounds are used in perfumes, creams, lipsticks, and other products to add depth to the fragrance; In the tobacco industry, fruit flavorings are primarily used to release sweet, fruity aromas that neutralize the harshness and bitterness of tobacco, mask irritation, modify the aftertaste, and enhance smoking comfort.

https://doi.org/10.3390/separations13060177

4. Conclusions

In this study, GC×GC-O-TOF-MS combined with a molecular sensory science approach was employed to separate and identify volatile aroma compounds in three natural fruit flavor extracts from plum, jujube, and grape. A total of 188 volatile compounds were identified, with 113, 103, and 85 compounds in plum, jujube, and grape extracts, respectively. Aroma extract dilution analysis revealed 14, 11, and 32 aroma-active compounds with flavor dilution factors ≥ 3, respectively. Quantification using internal standard calibration curves and calculation of odor activity values determined the relative contribution of each compound to the overall aroma. Octanal was identified as a key aroma-active compound shared by plum and jujube extracts, while ethyl cinnamate was common to both plum and grape extracts. Key aromatic compounds in plum extract, including angelica lactone, linalool, 5-methyl-2-furamethanol, methyl phenylacetate and ethyl cinnamate, contribute to its sweet flavor characteristics. In the jujube extract, 5-methyl-2-furfural, benzeneacetaldehyde, γ-nonanolactone, and ketoisophorone were the primary contributors to sweetness. In the grape extract, estragole, 4-methyl-2-pentyl-1,3-dioxolane, isovaleraldehyde propylene glycol acetal, and ethyl cinnamate were the main contributors to sweetness, while geranylacetone predominantly imparted fruity notes. A comprehensive two-dimensional gas chromatography–olfactometry–time-of-flight mass spectrometry (GC×GC–O–TOF–MS) method was established for the separation and identification of key aroma-active compounds in three complex natural fruit flavor extracts from plum, jujube, and grape. Through a molecular sensory science approach, this study deepens the molecular-level understanding of the aroma composition of these extracts, providing a theoretical basis for precise blending and flavor design of natural fragrances. The identified key aroma compounds can directly enhance the aroma quality of end products, facilitating the development of high-end fragrances, cosmetics, functional foods, and flavored beverages. However, flavor adjustments will still need to be made based on the product lineup. Furthermore, the specific processing methods for the three fruit-flavored natural flavor extracts are unknown, making it impossible to establish a link between the processing methods and the aromas, which presents certain limitations. Notwithstanding these limitations, the key aroma compounds identified here warrant further investigation into their biosynthetic pathways and regulatory mechanisms. Future research should focus on the biosynthetic pathways and regulatory mechanisms of these key aroma substances, promoting a systematic transition from structural elucidation to flavor innovation.

Author Contributions: Writing—original draft, R.C.; methodology and formal analysis, Q.J.; methodology and investigation, A.F. and F.Y.; methodology and investigation, Y.D.; investigation and formal analysis, J.S. and L.L.; supervision, writing—review and editing, Y.L.; supervision, writing—review and editing, S.B. All authors have read and agreed to the published version of the manuscript.

Funding: This research received no external funding.

Informed Consent Statement: All subjects underwent training in sensory evaluation techniques and descriptive quantitative analysis before the assessments. And all subjects signed an informed consent form, which was approved by the University Scientific Research Ethics Committee of BTBU (2025-50).

Data Availability Statement: The original contributions presented in this study are included in the article. Further inquiries can be directed to the corresponding author.

Conflicts of Interest: The authors declare no conflicts of interest.

https://doi.org/10.3390/separations13060177

Appendix A

Table A1. The volatile odor compounds in extracts of three fruit flavors: plum, jujube, and grape.

	Volatile Compounds	CAS	RI	Odor Description	Identification	Relative Concentration		
						Plum Extract	Jujube Extract	Grape Extract
Esters (45)								
1	Methyl acetate	79-20-9	-	sweet, fruity	MS/RI	4.948 ± 1.941	0.983 ± 0.5	-
2	Ethyl acetate	141-78-6	844	fruity, sweet	MS/RI	-	-	28.042 ± 24.599
3	Methyl propionate	554-12-1	865	fruity	MS/RI	-	0.013 ± 0.004	-
4	Methyl butyrate	623-42-7	951	fruity, sweet	MS/RI	-	0.041 ± 0.014	-
5	Butyl acetate	123-86-4	1038	fruity	MS/RI	-	0.006 ± 0.003	-
6	Butyl butyrate	109-21-7	1190	fruity	MS/RI	-	-	0.376 ± 0.48
7	Ethyl caproate	123-66-0	1205	sweet, fruity	MS/RI/O	0.227 ± 0.086	-	13.431 ± 4.537
8	Ethyl pyruvate	617-35-6	1229	fruity, sweet	MS/RI	-	-	0.536 ± 0.272
9	Hexyl acetate	142-92-7	1242	fruity, green	MS/RI	-	0.007 ± 0.005	0.399 ± 0.572
10	Isoamyl 2-methylbutyrate	27625-35-0	1268	sweet, fruity	MS/RI	0.003 ± 0.001	-	-
11	Ethyl heptanoate	106-30-9	1302	fruity	MS/RI	-	0.057 ± 0.049	1.346 ± 0.274
12	Methyl octanoate	111-11-5	1358	waxy, sweet	MS/RI	0.047 ± 0.039	0.23 ± 0.023	-
13	Butyl hexanoate	626-82-4	1384	fruity	MS/RI	-	0.001 ± 0	-
14	Angelica lactone	591-12-8	1390	sweet	MS/RI/O	0.119 ± 0.003	0.02 ± 0.008	-
15	Ethyl caprylate	106-32-1	1406	fruity	MS/RI	0.017 ± 0.007	-	5.304 ± 0.678
16	Ethyl ester-2,4-dimethyl-1,3-dioxolane-2-acetic acid	6290-17-1	1444	fruity	MS/RI	-	-	0.535 ± 0.812
17	Furfuryl acetate	623-17-6	1489	sweet, fruity	MS/RI	0.107 ± 0.013	0.012 ± 0.003	-
18	Ethyl nonanoate	123-29-5	1502	fruity, rose	MS/RI/O	-	-	11.517 ± 1.179
19	3,7-Dimethyl-1,6-octadien-3-ol formate	115-99-1	1503	citrus, green	MS/RI	-	0.046 ± 0.004	-
20	Bornyl acetate	76-49-3	1549	woody	MS/RI	-	-	0.145 ± 0.043
21	Methyl ester decanoic acid	110-42-9	1562	fruity, floral	MS/RI	-	0.107 ± 0.043	-
22	Methyl benzoate	93-58-3	1579	floral	MS/RI	0.246 ± 0.044	-	-
23	γ-Butyrolactone	96-48-0	1582	caramel	MS/RI	0.674 ± 0.137	-	-
24	Hexyl hexanoate	6378-65-0	1582	fruity, herbal	MS/RI/O	-	0.005 ± 0.002	0.282 ± 0.322
25	Ethyl caprate	110-38-3	1608	sweet, waxy	MS/RI	-	-	1.386 ± 0.069
26	Ethyl benzoate	93-89-0	1626	fruity	MS/RI/O	-	-	1.313 ± 0.197
27	Benzyl acetate	140-11-4	1682	sweet, floral	MS/RI	0.54 ± 0.027	0.048 ± 0.014	-
28	Ethyl ester undecanoic acid	627-90-7	1709	waxy	MS/RI	-	-	0.815 ± 0.106
29	Methyl phenylacetate	101-41-7	1711	sweet, floral	MS/RI/O	0.82 ± 0.047	-	-

https://doi.org/10.3390/separations13060177

Table A1. *Cont.*

	Volatile Compounds	CAS	RI	Odor Description	Identification	Relative Concentration		
						Plum Extract	Jujube Extract	Grape Extract
30	Ethyl 3-hydroxyhexanoate	2305-25-1	1722	fruity	MS/RI	-	-	0.393 ± 0.337
31	Methyl salicylate	119-36-8	1728	mint	MS/RI	1.433 ± 0.017	-	5.45 ± 4.959
32	Ethyl phenylacetate	101-97-3	1739	sweet, floral	MS/RI	-	-	0.749 ± 0.193
33	Diisopropyl adipate	6938-94-9	1829	sour	MS/RI	-	-	2.997 ± 4.92
34	5-Butyldihydro-2(3H)-furanone	104-50-7	1867	sweet, coconut	MS/RI	0.106 ± 0.012	-	-
35	γ-Nonanolactone	104-61-0	1975	sweet, waxy	MS/RI/O	-	0.856 ± 0.244	-
36	Isopropyl myristate	110-27-0	2003	fatty	MS/RI	-	0.035 ± 0.053	0.762 ± 1.228
37	δ-Nonalactone	3301-94-8	2028	sweet	MS/RI	0.05 ± 0.008	-	-
38	Ethyl cinnamate	103-36-6	2077	sweet, fruity	MS/RI/O	5.439 ± 0.887	-	3.521 ± 0.528
39	γ-Decalactone	706-14-9	2092	waxy	MS/RI/O	-	0.212 ± 0.029	0.371 ± 0.209
40	Ethylpentadecanoate	41114-00-5	2118	sweet	MS/RI/O	-	-	0.218 ± 0.175
41	Ethyl palmitate	628-97-7	2216	fruity, waxy	MS/RI/O	0.07 ± 0.032	-	34.961 ± 7.252
42	5,6,7,7a-Tetrahydro-4,4,7a-trimethyl-2(4H)-benzofuranone	15356-74-8	2286	fruit, woody	MS/RI	0.44 ± 0.045	0.133 ± 0.055	-
43	Triethyl citrate	77-93-0	2395	fruity, wine	MS/RI	-	0.049 ± 0.023	-
44	Benzyl Benzoate	120-51-4	-	sweet	MS/RI	11.748 ± 0.732	0.038 ± 0.03	-
45	Ethyl formate	109-94-4	-	green, rose	MS/RI	-	-	5.272 ± 1.173
Aldehydes (33)								
1	Acetaldehyde	75-07-0	-	fruity	MS/RI	0.125 ± 0.051	0.04 ± 0.021	0.752 ± 0.729
2	Glutaraldehyde	111-30-8	830	fruity	MS/RI	0.626 ± 0.051	0.012 ± 0.007	-
3	Butyraldehyde	123-72-8	831	cocoa, musty	MS/RI	0.023 ± 0.001	-	2.065 ± 2.889
4	3-Methylbutyraldehyde	590-86-3	879	chocolate	MS/RI	2.005 ± 0.481	0.991 ± 0.553	6.424 ± 0.844
5	4-Methyl-2-pentyl-1,3-dioxolane	1599-49-1	1041	fruity, waxy	MS/RI/O	5.181 ± 0.97	-	2.063 ± 3.443
6	Hexanal	66-25-1	1047	fruity	MS/RI/O	0.048 ± 0.045	0.267 ± 0.116	8.017 ± 2.093
7	(*E*)-2-Methyl-2-butenal	497-03-0	1059	green, fruit	MS/RI	0.048 ± 0.004	0.015 ± 0.004	-
8	Isovaleraldehyde propylene glycol acetal	18433-93-7	1140	sweet, fruity	MS/RI/O	0.388 ± 0.062	-	3.904 ± 0.418
9	Heptanal	111-71-7	1151	green, herbal	MS/RI	-	0.105 ± 0.003	-
10	3-Methyl-2-butenal	107-86-8	1162	sweet, fruity	MS/RI	25.597 ± 2.891	0.002 ± 0.001	-
11	Octanal	124-13-0	1256	waxy	MS/RI/O	2.005 ± 0.481	0.199 ± 0.096	3.403 ± 0.986
12	Nonanal	124-19-6	1362	waxy, rose	MS/RI	0.808 ± 0.146	0.724 ± 0.492	-
13	(*E*)-2-Octenal	2548-87-0	1395	green, waxy	MS/RI/O	4.681 ± 0.088	-	2.763 ± 0.335
14	2-Heptyl-3-methyl-1,3-Dioxolane	74094-61-4	1397	green, woody	MS/RI	2.298 ± 0.513	-	1.973 ± 3.287
15	Furfural	98-01-1	1421	sweet, woody	MS/RI/O	0.048 ± 0.004	0.283 ± 0.011	-
16	Decanal	112-31-2	1464	sweet, waxy	MS/RI/O	2.357 ± 0.426	0.194 ± 0.029	2.984 ± 2.029
17	Undecanal	112-44-7	1472	waxy	MS/RI/O	-	0.034 ± 0.011	0.164 ± 0.071
18	Benzaldehyde	100-52-7	1478	sweet	MS/RI/O	-	3.158 ± 0.198	18.584 ± 31.923

https://doi.org/10.3390/separations13060177

Table A1. *Cont.*

	Volatile Compounds	CAS	RI	Odor Description	Identification	Relative Concentration		
						Plum Extract	Jujube Extract	Grape Extract
19	5-Methyl-2-furaldehyde	620-02-0	1529	sweet, caramel	MS/RI/O	111.099 ± 65.359	7.84 ± 13.565	-
20	Pyrrole-2-carboxaldehyde	1192-58-1	1576	roasted, nutty	MS/RI/O	-	0.85 ± 0.31	-
21	α,4-Dimethyl-3-cyclohexene-1-acetaldehyde	29548-14-9	1579	herbal	MS/RI	0.808 ± 0.146	-	-
22	2,6,6-Trimethyl-1-cyclohexene-1-carboxaldehyde	432-25-7	1586	herbal	MS/RI	-	0.011 ± 0.001	0.471 ± 0.218
23	Benzeneacetaldehyde	122-78-1	1596	sweet, floral	MS/RI/O	-	0.132 ± 0.049	-
24	2,3-Dihydro-2,2,6-trimethylben	116-26-7	1608	herbal	MS/RI	-	0.157 ± 0.026	-
25	Dodecanal	112-54-9	1676	waxy	MS/RI/O	-	-	0.459 ± 0.333
26	2,4-Dimethyl-benzaldehyde	15764-16-6	1765	vanilla	MS/RI	-	-	-
27	4-Methyl-2-phenyl-1,3-dioxolane	2568-25-4	1798	sweet	MS/RI	-	-	21.32 ± 30.925
28	2-Methyl-3(2-furyl)acrolein	874-66-8	1824	woody	MS/RI	-	0.086 ± 0.022	-
29	Pyrrole-2-carboxaldehyde	1003-29-8	1964	musty	MS/RI/O	-	0.543 ± 0.037	-
30	2-(Phenylmethylene)-octanal	101-86-0	2301	floral, green	MS/RI	2.357 ± 0.426	-	-
31	5-Hydroxymethylfurfural	67-47-0	-	fatty, waxy	MS/RI	40.559 ± 3.088	1.43 ± 0.348	7.117 ± 11.176
32	Propanal	123-38-6	-	earthy	MS/RI	-	0.014 ± 0.004	-
33	2-Phenyl-3-(2-furyl)-propenal	57568-60-2	-	spice	MS/RI	-	0.011 ± 0.002	-
Alcohols (29)								
1	2-Methyl-2-propanol	75-65-0	861	camphor	MS/RI/O	7.902 ± 11.067	0.233 ± 0.185	2.082 ± 0.408
2	Ethanol	64-17-5	899	alcoholic	MS/RI/O	0.39 ± 0.011	0.092 ± 0.058	0.882 ± 1.362
3	1-Propanol	71-23-8	1007	alcoholic, fermented	MS/RI	0.414 ± 0.161	-	-
4	2-Methyl-3-buten-2-ol	115-18-4	1008	herbal, earthy	MS/RI	0.005 ± 0.004	-	0.089 ± 0.077
5	2-Methyl-1-propanol	78-83-1	1057	ethereal, winey	MS/RI/O	0.088 ± 0.035	0.035 ± 0.007	-
6	2-Propen-1-ol	107-18-6	1075	mustard	MS/RI	0.092 ± 0.01	-	-
7	1-Butanol	71-36-3	1110	sweet	MS/RI	0.148 ± 0.019	0.022 ± 0.008	1.341 ± 1.66
8	3-Methyl-1-butanol	123-51-3	1173	fruity	MS/RI	0.139 ± 0.021	-	-
9	1-Pentanol	71-41-0	1215	sweet	MS/RI	1.037 ± 0.025	0.347 ± 0.117	-
10	Prenol	556-82-1	1283	fruity	MS/RI	0.007 ± 0.001	0.004 ± 0.002	-
11	1-Hexanol	111-27-3	1316	fruity	MS/RI	6.661 ± 0.326	-	-
12	5-Hexen-1-ol	821-41-0	1372	green	MS/RI	0.174 ± 0.003	-	1.101 ± 0.41
13	1-Hexanol	5989-33-3	1411	floral, sweet	MS/RI	0.584 ± 0.09	0.388 ± 0.136	22.81 ± 26.51
14	1-Octen-3-ol	3391-86-4	1413	earthy, mushroom	MS/RI	1.465 ± 0.016	-	1.037 ± 0.171
15	2-Ethylhexanol	104-76-7	1450	citrus, floral	MS/RI	4.55 ± 0.604	1.275 ± 0.365	-
16	Linalool	78-70-6	1503	floral, sweet	MS/RI/O	0.588 ± 0.054	0.062 ± 0.044	2.834 ± 0.746
17	1-Octanol	111-87-5	1516	waxy, green	MS/RI	2.946 ± 0.224	-	1.574 ± 0.983
18	Isophorone	78-59-1	1556	woody, sweet	MS/RI	0.395 ± 0.088	0.181 ± 0.058	-
19	2-Furanmethanol	98-00-0	1612	sweet, caramel	MS/RI	0.065 ± 0.017	-	-

https://doi.org/10.3390/separations13060177

Table A1. *Cont.*

	Volatile Compounds	CAS	RI	Odor Description	Identification	Relative Concentration		
						Plum Extract	Jujube Extract	Grape Extract
20	1-Nonanol	143-08-8	1620	floral	MS/RI/O	0.021 ± 0.004	-	-
21	α,α,4-Trimethyl-(R)-3-cyclohexene-1-methanol	7785-53-7	1657	floral	MS/RI	0.177 ± 0.078	-	-
22	Borneol	507-70-0	1661	woody, camphor	MS/RI	0.138 ± 0.131	-	5.631 ± 1.736
23	5-Methyl-2-furanmethanol	3857-25-8	1672	sweet, caramellic	MS/RI/O	10.05 ± 1.718	0.767 ± 0.645	4.13 ± 5.795
24	(3R,6S)-2,2,6-Trimethyl-6-vinyltetrahydro-2H-pyran-3-ol	39028-58-5	1717	woody	MS/RI	1.227 ± 0.047	0.516 ± 0.111	-
25	1-Decanol	112-30-1	1719	waxy, floral	MS/RI	-	-	0.302 ± 0.093
26	2,5-Dimethyl-2,5-hexanediol	110-03-2	1720	camphor	MS/RI	-	-	1.241 ± 0.292
27	1,1′-Oxybis-2-propanol	110-98-5	1782	alcoholic	MS/RI	-	-	4.724 ± 1.849
28	Benzyl alcohol	100-51-6	1824	floral, rose	MS/RI/O	-	0.206 ± 0.054	0.346 ± 0.14
29	Phenylethyl alcohol	60-12-8	1861	floral, rose	MS/RI/O	-	0.021 ± 0.006	-
Ketones (29)								
1	2-Butanone	78-93-3	858	fruity	MS/RI	-	0.074 ± 0.013	3.347 ± 2.173
2	2,3-Butanedione	431-03-8	938	sweet	MS/RI/O	-	0.619 ± 0.209	1.274 ± 0.581
3	2,3-Pentanedione	600-14-6	1022	sweet	MS/RI	0.739 ± 0.111	0.215 ± 0.042	-
4	2-Heptanone	110-43-0	1149	fruity, sweet	MS/RI	-	0.017 ± 0.006	-
5	6-Methyl-2-heptanone	928-68-7	1207	camphoreous	MS/RI	0.006 ± 0.004	-	0.138 ± 0.134
6	5-Methyl-3-heptanone	541-85-5	1224	herbal, sweet	MS/RI	-	-	0.316 ± 0.123
7	Dihydro-2-methyl-3(2H)-furanone	3188-00-9	1226	sweet	MS/RI	1.328 ± 0.538	0.049 ± 0.014	0.667 ± 0.845
8	Acetoin	513-86-0	1245	sweet	MS/RI	0.029 ± 0.009	0.042 ± 0.007	0.98 ± 0.192
9	2-Octanone	111-13-7	1253	earthy, weedy	MS/RI	0.024 ± 0.006	0.02 ± 0.002	-
10	1-Hydroxy-2-propanone	116-09-6	1258	sweet	MS/RI	0.177 ± 0.07	0.107 ± 0.045	-
11	6-Methyl-5-hepten-2-one	110-93-0	1301	citrus, green	MS/RI	0.132 ± 0.02	0.043 ± 0.03	13.882 ± 2.407
12	2-Nonanone	821-55-6	1357	sweet, green	MS/RI	0.013 ± 0.007	-	0.372 ± 0.305
13	2-Decanone	693-54-9	1459	floral	MS/RI/O	-	-	1.043 ± 0.191
14	Cis-5-methyl-2-(1-methylethyl)-cyclohexanone	491-07-6	1459	sweet	MS/RI	-	-	0.284 ± 0.139
15	6-Methyl-3,5-heptadiene-2-one	1604-28-0	1550	woody, sweet	MS/RI	0.183 ± 0.01	-	-
16	2-Undecanone	112-12-9	1565	waxy, fruity	MS/RI	-	-	0.163 ± 0.014
17	Acetophenone	98-86-2	1608	sweet	MS/RI	-	0.135 ± 0.015	13.288 ± 8.505
18	2-Hydroxy-3,5,5-trimethyl-cyclohex-2-enone	4883-60-7	1622	woody	MS/RI	0.012 ± 0.003	-	-
19	Ketoisophorone	1125-21-9	1648	woody, sweet	MS/RI/O	-	0.044 ± 0.001	-
20	6,10-Dimethyl-2-undecanone	1604-34-8	1654	musty	MS/RI	-	-	0.652 ± 0.198
21	2,6,6-Trimethyl-2,4-cycloheptadien-1-one	503-93-5	1689	minty	MS/RI	0.06 ± 0.014	-	-
22	2(5H)-Furanone	497-23-4	1699	buttery	MS/RI	0.157 ± 0.071	0.123 ± 0.043	-
23	α-Ionone	127-41-3	1810	sweet, woody	MS/RI	0.054 ± 0.005	-	-
24	Geranylacetone	3796-70-1	1814	fruity, waxy	MS/RI/O	-	-	4.416 ± 7.54

https://doi.org/10.3390/separations13060177

Table A1. *Cont.*

	Volatile Compounds	CAS	RI	Odor Description	Identification	Relative Concentration		
						Plum Extract	Jujube Extract	Grape Extract
25	4-(2-Furanyl)-3-buten-2-one	623-15-4	1853	sweet, spicy	MS/RI	0.165 ± 0.059	-	-
26	trans-β-Ionone	79-77-6	1898	floral, woody	MS/RI	-	0.002 ± 0.001	-
27	6,10,14-Trimethyl-2-pentadecanone	502-69-2	2090	herbal	MS/RI	-	-	0.873 ± 0.986
28	5,6-Dihydro-6-pentyl-2H-pyran-2-one	54814-64-1	2176	sweet	MS/RI	-	0.023 ± 0.02	-
29	Acetone	67-64-1	-	fruity	MS/RI/O	-	0.89 ± 0.526	5.861 ± 1.452
Heterocyclic (22)								
1	Furan	110-00-9	-	ethereal	MS/RI	0.189 ± 0.192	-	-
2	2,5-Dimethylfuran	625-86-5	919	ethereal	MS/RI	-	0.021 ± 0.008	-
3	2-Ethyl-5-methyl-Furan	1703-52-2	1003	fresh	MS/RI	-	0.003 ± 0.003	-
4	2-Vinylfuran	1487-18-9	1037	coffee	MS/RI	-	0.014 ± 0.007	-
5	2-Pentyl-furan	3777-69-3	1202	fruity green	MS/RI	0.035 ± 0.004	0.015 ± 0.003	3.847 ± 2.85
6	Thiazole	288-47-1	1212	nutty	MS/RI	-	0.004 ± 0.002	-
7	Methyl-pyrazine	109-08-0	1230	nutty	MS/RI	0.04 ± 0.004	0.006 ± 0.002	-
8	2,6-Dimethyl-pyrazine	108-50-9	1293	cocoa	MS/RI	2.055 ± 0.261	0.197 ± 0.043	-
9	Ethyl-pyrazine	13925-00-3	1297	nutty	MS/RI	-	0.005 ± 0.003	-
10	2,3-Dimethyl-pyrazine	5910-89-4	1310	nutty	MS/RI	0.005 ± 0	0.004 ± 0.001	-
11	2-Ethyl-6-methyl-pyrazine	13925-03-6	1351	roasted potato	MS/RI	0.134 ± 0.004	-	-
12	Trimethyl-pyrazine	14667-55-1	1372	nutty	MS/RI	0.057 ± 0.016	0.128 ± 0.035	-
13	2,3-Dimethyl-5-ethylpyrazine	15707-34-3	1427	popcorn	MS/RI	-	0.043 ± 0.013	-
14	Tetramethyl-pyrazine	1124-11-4	1441	nutty	MS/RI	-	0.005 ± 0.001	-
15	2-Acetylfuran	1192-62-7	1459	sweet	MS/RI/O	17.64 ± 24.385	7.135 ± 0.897	11.774 ± 10.526
16	1-(2-Furanyl)-1-propanone	3194-15-8	1529	fruity	MS/RI	-	3.699 ± 0.294	-
17	2-Methyl-benzofuran	4265-25-2	1553	burnt	MS/RI	0.057 ± 0.006	-	-
18	Pyrrole-2-carboxaldehyde	2167-14-8	1566	burnt	MS/RI	4.438 ± 0.045	13.875 ± 1.34	-
19	2-Acetyl-5-methylfuran	1193-79-9	1570	nutty	MS/RI	2.795 ± 0.23	0.102 ± 0.169	-
20	Methyl 6-methyl-2-pyridyl ketone	6940-57-4	1588	potato	MS/RI	0.792 ± 0.009	0.003 ± 0.001	-
21	1-(5-Methyl-2-furanyl)-1-propanone	10599-69-6	1639	green	MS/RI	0.339 ± 0.039	-	-
22	1-(1H-pyrrol-2-yl)-ethanone	1072-83-9	1915	fruity	MS/RI/O	21.679 ± 2.327	0.479 ± 0.718	-
Acids (7)								
1	Acetic acid	64-19-7	1407	sour, vinegar	MS/RI/O	21.791 ± 0.003	8.444 ± 14.536	22.518 ± 25.017
2	2-Methyl-propanoic acid	79-31-2	1534	sour, cheese	MS/RI	0.651 ± 0.114	-	-
3	Hexanoic acid	142-62-1	1810	sour	MS/RI	3.339 ± 0.294	-	15.817 ± 7.939
4	Heptanoic acid	111-14-8	1918	sour	MS/RI/O	0.841 ± 0.561	-	1.518 ± 1.537
5	Octanoic acid	124-07-2	2013	waxy	MS/RI/O	4.469 ± 1.331	-	7.476 ± 4.187
6	Nonanoic acid	112-05-0	2114	waxy	MS/RI/O	-	6.706 ± 5.638	34.427 ± 8.719

https://doi.org/10.3390/separations13060177

Table A1. *Cont.*

	Volatile Compounds	CAS	RI	Odor Description	Identification	Relative Concentration		
						Plum Extract	Jujube Extract	Grape Extract
7	Benzoic acid	65-85-0	2392	balsam	MS/RI	-	-	12.381 ± 11.056
Phenols (6)								
1	2-Methoxy-phenol	90-05-1	1807	woody	MS/RI/O	0.125 ± 0.051	-	-
2	Butylated Hydroxytoluene	128-37-0	1869	camphor	MS/RI	0.626 ± 0.051	-	1.101 ± 0.567
3	Eugenol	97-53-0	2107	sweet, woody	MS/RI/O	0.023 ± 0.001	-	-
4	2-Methoxy-4-vinylphenol	7786-61-0	2133	sweet woody	MS/RI	2.005 ± 0.481	0.139 ± 0.053	-
5	2,4-Di-tert-butylphenol	96-76-4	2244	phenolic	MS/RI	5.181 ± 0.97	6.885 ± 0.785	-
6	(Z)-2-Methoxy-4-(1-propenyl)-phenol	5912-86-7	2278	clove	MS/RI	0.048 ± 0.045	-	-
Olefins (5)								
1	(-)-Limonene	5989-54-8	1174	herbal	MS/RI	0.152 ± 0.005	-	-
2	(4R)-1-Methyl-4-(1-methylethenyl)-cyclohexene	5989-27-5	1174	citrus	MS/RI	-	0.274 ± 0.218	52.802 ± 14.743
3	Styrene	100-42-5	1222	sweet	MS/RI	0.116 ± 0.092	3.288 ± 5.633	-
4	(Z)-1-Methoxy-4-(1-propenyl)-benzene	25679-28-1	1781	anise	MS/RI	-	0.028 ± 0.019	-
5	(-)-Calamenene	483-77-2	1801	herbal	MS/RI	-	-	0.506 ± 0.352
Others (11)								
1	Dimethyl disulfide	624-92-0	1037	onion	MS/RI	0.626 ± 0.051	0.004 ± 0.002	-
2	2-Butyl-4-methyl-1,3-dioxolane	74094-60-3	1136	nutty	MS/RI	0.023 ± 0.001	0.112 ± 0.184	-
3	Trimethyl-oxazol	20662-84-4	1165	nutty	MS/RI	2.005 ± 0.481	0.005 ± 0.002	-
4	2-(Methoxymethyl)-furan	13679-46-4	1202	coffee	MS/RI	5.181 ± 0.97	-	-
5	p-Cymene	99-87-6	1238	woody	MS/RI	0.048 ± 0.045	-	0.781 ± 0.27
6	Dimethyl trisulfide	3658-80-8	1344	onion	MS/RI	0.048 ± 0.004	0.003 ± 0.001	-
7	trans-Linalool oxide (furanoid)	34995-77-2	1436	floral	MS/RI	0.388 ± 0.062	-	-
8	(-)-Camphor	464-48-2	1479	camphor	MS/RI	25.597 ± 2.891	0.007 ± 0.006	-
9	Dill ether	74410-10-9	1482	dill	MS/RI	111.099 ± 65.359	-	-
10	Estragole	140-67-0	1783	sweet	MS/RI/O	1.324 ± 0.092	-	2.014 ± 3.141
11	Heptane	142-82-5	-	sweet	MS/RI	0.808 ± 0.146	-	1.224 ± 2.058

-: not detected; MS: mass spectral data; RI: retention index, the actual RI value calculated from the peak time of the series of alkanes on the DB-WAX column; O: sniff.

https://doi.org/10.3390/separations13060177

Table A2. The aroma-active compounds in extracts of three fruit flavors: plum, jujube, and grape.

	Volatile Compounds	CAS	Odor Description	FD Factor		
				Plum Extract	Jujube Extract	Grape Extract
Aldehydes (15)						
1	4-Methyl-2-pentyl-1,3-dioxolane	1599-49-1	fruity, waxy	-	-	3
2	Hexanal	66-25-1	fruity	-	-	1
3	Isovaleraldehyde propylene glycol acetal	18433-93-7	sweet, fruity	-	-	729
4	Octanal	124-13-0	waxy	3	3	3
5	(*E*)-2-Octenal	2548-87-0	green, waxy	-	-	1
6	Furfural	98-01-1	sweet, woody	3	1	-
7	Decanal	112-31-2	sweet, waxy	-	-	1
8	Undecanal	112-44-7	waxy	-	-	1
9	Benzaldehyde	100-52-7	sweet	3	-	-
10	5-Methyl-2-furaldehyde	620-02-0	caramel	3	-	-
11	Pyrrole-2-carboxaldehyde	1192-58-1	roasted, nutty	-	1	-
12	Benzeneacetaldehyde	122-78-1	sweet, floral	9	9	-
13	2,3-Dihydro-2,2,6-trimethylben	116-26-7	herbal	-	-	-
14	Dodecanal	112-54-9	waxy	-	-	1
15	Pyrrole-2-carboxaldehyde	1003-29-8	musty	27	-	-
Alcohols (11)						
1	2-Methyl-2-propanol	75-65-0	camphor	-	-	1
2	Ethanol	64-17-5	alcoholic	-	-	1
3	2-Methyl-3-buten-2-ol	115-18-4	herbal, earthy	-	-	-
4	2-Methyl-1-propanol	78-83-1	ethereal, winey	-	-	1
5	1-Hexanol	111-27-3	fruity	-	1	-
6	1-Octen-3-ol	3391-86-4	earthy, mushroom	-	-	9
7	Linalool	78-70-6	floral, sweet	9	-	-
8	1-Nonanol	143-08-8	floral	-	-	1
9	5-Methyl-2-furanmethanol	3857-25-8	sweet, caramellic	9	3	-
10	Benzyl alcohol	100-51-6	floral, rose	-	-	1
11	Phenylethyl alcohol	60-12-8	floral, rose	9	1	-
Esters (11)						
1	Ethyl caproate	123-66-0	sweet, fruity	-	-	3
2	Angelica lactone	591-12-8	sweet	3	-	-
3	Ethyl nonanoate	123-29-5	fruity, rose	-	-	1
4	Hexyl hexanoate	6378-65-0	fruity, herbal	-	-	1

https://doi.org/10.3390/separations13060177

Table A2. *Cont.*

	Volatile Compounds	CAS	Odor Description	FD Factor		
				Plum Extract	Jujube Extract	Grape Extract
5	Ethyl benzoate	93-89-0	fruity	-	-	1
6	Methyl phenylacetate	101-41-7	sweet, floral	3	-	-
7	Ethyl cinnamate	103-36-6	sweet, fruity	729	-	729
8	γ-Decalactone	706-14-9	waxy	-	27	-
9	Ethylpentadecanoate	41114-00-5	sweet	-	-	1
10	Ethyl palmitate	628-97-7	waxy, fruity	-	-	1
11	γ-Nonanolactone	104-61-0	coconut, waxy	-	9	-
Ketones (5)						
1	2,3-Butanedione	431-03-8	sweet	-	-	1
2	2-Decanone	693-54-9	floral	-	-	1
3	Ketoisophorone	1125-21-9	woody, sweet	-	3	-
4	Geranylacetone	3796-70-1	fruity, waxy	-	-	3
5	Acetone	67-64-1	fruity	-	-	1
Acids (4)						
1	Acetic acid	64-19-7	sour, vinegar	-	-	1
2	Heptanoic acid	111-14-8	sour	-	-	1
3	Octanoic acid	124-07-2	waxy	-	-	1
4	Nonanoic acid	112-05-0	waxy	-	-	1
Phenols (2)						
1	2-Methoxy-phenol	90-05-1	woody	243	-	-
2	Eugenol	97-53-0	sweet, woody	243	-	-
Heterocyclic (2)						
1	2-Acetylfuran	1192-62-7	sweet	-	-	1
2	1-(1H-pyrrol-2-yl)-ethanone	1072-83-9	fruity	-	1	243
Others (1)						
1	Estragole	140-67-0	sweet	-	-	729

-: not detected.

https://doi.org/10.3390/separations13060177

Table A3. Concentrations, odor thresholds, and odor activity values of aroma-active compounds in extracts of three fruit flavors: plum, grape, and jujube.

	Aroma-Active Compounds	Linear Equations	R^3	Linear Range (μg/L)	Concentration (μg/g)			LOD	LOQ	RSD (%)	Odor Threshold (ug/g)	OAV		
					Plum Extract	Jujube Extract	Grape Extract					Plum Ex-tract	Jujube Ex-tract	Grape Extract
Alcohols														
1	1-Octen-3-ol	y = 0.3245x − 0.000005	0.9969	130.731~8370.000	-	-	1882.864 ± 238.500	0.00066	0.0020	12.7	1.5	-	-	1255
2	Linalool	y = 0.0868x + 0.000006	0.9989	27.125~868.000	672.494 ± 73.571	-	-	0.000048	0.00015	10.9	6	112	-	-
3	5-Methyl-2-furanmethanol *	y = 0.1619x − 0.000002	0.9984	33.653~538.450	147.397 ± 18.844	351.631 ± 30.605	-	0.000029	0.000088	12.8	0.009	16,749	39,958	-
4	Phenethyl alcohol	y = 0.0595x − 0.00001	0.9918	637.500~10,200.00	986.834 ± 59.631	-	-	0.001099	0.0034	6.0	564	2	-	-
Aldehydes														
1	Nonanal	y = 1.9639x + 0.0000008	0.9974	12.800~411.000	17.917 ± 1.327	90.921 ± 6.825	72.586 ± 10.384	0.0000016	0.0000050	14.3	0.587	31	155	124
2	Furfural	y = 0.2732x + 0.0141	0.9907	36,250.000~1,160,000	297,076.482 ± 24,444.384	-	-	0.14	0.42	0.8	9562	31	-	-
3	Benzaldehyde	y = 0.9449x + 0.0003	0.9979	325.000~10,400.000	1815.849 ± 218.245	-	-	0.00060	0.0018	12.0	750.89	2	-	-
4	5-Methyl-2-furfural	y = 0.4509x − 0.004	0.9979	34,593.750~553,500.000	47,857.922 ± 5856.231	-	-	0.034	0.10	12.2	1110	43	-	-
5	Benzeneacetaldehyde	y = 0.8886x − 0.00003	0.9928	160.469~10,270.000	521.623 ± 63.830	440.280 ± 46.822	-	0.0011	0.0033	12.2	6.300	83	70	-
6	Pyrrole-2-carboxaldehyde	y = 0.2465x − 0.00006	0.9913	187.031~5985.000	2468.591 ± 225.528	-	-	0.00070	0.0021	9.1	65	38	-	-
7	4-Methyl-2-pentyl-1,3-dioxolanel *	y = 0.5658x + 0.00003	0.9949	140.672~9003.000	-	-	3210.760 ± 259.337	0.00040	0.0012	8.1	0.00902	-	-	355,960
8	Isovaleraldehyde propylene glycol acetal *	y = 0.7832x + 0.00005	0.994	279.688~17,900.000	-	-	9958.683 ± 700.445	0.0027	0.0081	7.0	0.000173	-	-	57,564,641
Ketones														
1	Ketoisophorone	y = 0.3044x + 0.000006	0.9990	40.692~1302.149	-	146.435 ± 16.183	-	0.000045	0.00014	11.1	25	-	6	-
2	Geranylacetone	y = 0.4208x − 0.00003	0.9953	272.813~4365.000	-	-	2064.986 ± 169.471	0.00040	0.0012	8.2	60	-	-	34
Esters														
1	γ-Decalactone	y = 0.6568x + 0.000006	0.9942	14.813~474.000	-	65.479 ± 8.303	-	0.000065	0.00020	12.7	1.100	-	60	-
2	Angelica lactone *	y = 0.1132x − 0.00005	0.9900	17.516~1121.000	454.138 ± 36.520	-	-	0.00015	0.00047	8.0	0.0875	5190	-	-
3	Methyl phenylacetate *	y = 0.4044x + 0.000003	0.9973	10.410~666.250	221.712 ± 22.081	-	-	0.000041	0.00012	10.0	0.000441	502,747	-	-

https://doi.org/10.3390/separations13060177

Table A3. *Cont.*

	Aroma-Active Compounds	Linear Equations	R^3	Linear Range (μg/L)	Concentration (μg/g)			LOD	LOQ	RSD (%)	Odor Threshold (ug/g)	OAV		
					Plum Extract	Jujube Extract	Grape Extract					Plum Ex-tract	Jujube Ex-tract	Grape Extract
4	Ethyl cinnamate	y = 0.506x − 0.00005	0.9991	20.860~1056.000	55.293 ± 8.569	-	728.362 ± 160.313	0.000041	0.00012	15.5	1.45	38	-	502
5	Ethyl caproate	y = 0.1349x + 0.0000004	0.9947	271.563~4345.000	-	-	2225.610 ± 181.657	0.0000052	0.00005	8.2	5	-	-	445
6	γ-Nonanolactone	y = 0.4363x − 0.00000006	0.9902	30.500~488.000	-	289.821 ± 39.777	-	0.000049	0.00015	13.7	9.700	-	30	-
Phenols														
1	Eugenol	y = 0.0645x + 0.0000003	0.9952	10.420~666.875	28.572 ± 4.464	-	-	0.000033	0.00010	15.6	0.61	47	-	-
2	Guaiacol	y = 0.231x − 0.000006	0.9919	89.207~1427.307	542.220 ± 72.182	-	-	0.00014	0.00041	13.3	10	54	-	-
Others														
1	Artemisia ketone	y = 0.6217x − 0.00009	0.9919	301.563~4825.000	-	-	337.489 ± 31.578	0.00058	0.0018	9.4	16	-	-	21

-: not found; Odor thresholds of each compound were referenced from a book named: Odour Thresholds: Compilations of Odour Threshold Values in Air, Water and Other Media. The odor thresholds for isovaleraldehyde propylene glycol acetal, methyl phenylacetate, 4-methyl-2-pentyl-1,3-dioxolane, angelic lactone, and 5-methyl-2-furanmethanol were determined through threshold experiments; *: the threshold values of these compounds were determined through threshold experiments.

https://doi.org/10.3390/separations13060177

References

1. Sharmeen, J.B.; Mahomoodally, F.M.; Zengin, G.; Maggi, F. Essential Oils as Natural Sources of Fragrance Compounds for Cosmetics and Cosmeceuticals. *Molecules* **2021**, *26*, 666. [CrossRef] [PubMed]
2. Bailly, C. Anticancer properties of *Prunus mume* extracts (Chinese plum, Japanese apricot). *J. Ethnopharmacol.* **2020**, *246*, 112215. [CrossRef] [PubMed]
3. Bendaali, Y.; Vaquero, C.; Escott, C.; González, C.; Morata, A. Isotonic Drinks Based on Organic Grape Juice and Naturally Flavored with Herb and Spice Extracts. *Beverages* **2023**, *9*, 49. [CrossRef]
4. Serafini, M.; Peluso, I. Functional Foods for Health: The Interrelated Antioxidant and Anti-Inflammatory Role of Fruits, Vegetables, Herbs, Spices and Cocoa in Humans. *Curr. Pharm. Des.* **2016**, *22*, 6701–6715. [CrossRef] [PubMed]
5. Prebihalo, S.E.; Berrier, K.L.; Freye, C.E.; Bahaghighat, H.D.; Moore, N.R.; Pinkerton, D.K.; Synovec, R.E. Multidimensional Gas Chromatography: Advances in Instrumentation, Chemometrics, and Applications. *Anal. Chem.* **2018**, *90*, 505–532. [CrossRef] [PubMed]
6. Hu, X.F.; Li, D.Y.; Ding, Y.; Zhang, Y.B.; Ren, C.G. Characteristic of volatile flavor compounds in 'Fengtangli' plum (*Prunus salicina* Lindl.) were explored based on GCxGC-TOF MS. *Front. Nutr.* **2025**, *12*, 1536954. [CrossRef] [PubMed]
7. Liu, H.C.; Yu, Y.S.; Zou, B.; Yu, Y.Y.; Yang, J.G.; Xu, Y.J.; Chen, X.W.; Yang, F. Evaluation of Dynamic Changes and Regularity of Volatile Flavor Compounds for Different Green Plum (*Prunus mume* Sieb. et Zucc) Varieties during the Ripening Process by HS-GC-IMS with PLS-DA. *Foods* **2023**, *12*, 551. [CrossRef] [PubMed]
8. Yue, R.R.; Zhang, Z.; Shi, Q.Q.; Duan, X.S.; Wen, C.P.; Shen, B.Q.; Li, X.A. Identification of the key genes contributing to the LOX-HPL volatile aldehyde biosynthesis pathway in jujube fruit. *Int. J. Biol. Macromol.* **2022**, *222*, 285–294. [CrossRef] [PubMed]
9. Liu, R.J.; Ma, L.; Meng, X.Y.; Zhang, S.W.; Cao, M.; Kong, D.C.; Chen, X.X.; Li, Z.Q.; Pang, X.M.; Bo, W.H. Volatile Profile Characterization of Jujube Fruit via HS-SPME-GC/MS and Sensory Evaluation. *Plants* **2024**, *13*, 1517. [CrossRef] [PubMed]
10. Wu, Y.S.; Zhang, W.W.; Yu, W.J.; Zhao, L.P.; Song, S.R.; Xu, W.P.; Zhang, C.X.; Ma, C.; Wang, L.; Wang, S.P. Study on the volatile composition of *table grapes* of three aroma types. *Lwt-Food Sci. Technol.* **2019**, *115*, 108450. [CrossRef]
11. Wang, J.Y.; Zhang, C.Y.; Li, M.; Fan, X.C.; Zhang, Y.; Jiang, J.F.; Liu, C.H.; Sun, L. Multi-omics analysis unravels the dynamics of flavor profile in 'Queen Nina' berries and the regulatory mechanism underlying volatile esters. *Food Chem. Mol. Sci.* **2025**, *10*, 100248. [CrossRef] [PubMed]
12. Yang, X.Y.; Zeng, P.P.; Wen, J.; Wang, C.L.; Yao, L.Y.; He, M. Gain deeper insights into traditional Chinese medicines using multidimensional chromatography combined with chemometric approaches. *Chin. Herb. Med.* **2024**, *16*, 27–41. [CrossRef] [PubMed]
13. Dash, M.; Singh, S.; Sahoo, S.; Kar, B. Targeted Analyte Deconvolution and Detection of *Piper longum* Leaf Volatiles Using Two-Dimensional Gas Chromatography and Time-of-Flight Mass Spectrometry. *Anal. Lett.* **2024**, *57*, 1462–1473. [CrossRef]
14. Tranchida, P.Q.; Sciarrone, D.; Dugo, P.; Mondello, L. Heart-cutting multidimensional gas chromatography: A review of recent evolution, applications, and future prospects. *Anal. Chim. Acta* **2012**, *716*, 66–75. [CrossRef] [PubMed]
15. Gou, M.; Bi, J.F.; Chen, Q.Q.; Wu, X.Y.; Fauconnier, M.L.; Qiao, Y.N. Advances and Perspectives in Fruits and Vegetables Flavor Based on Molecular Sensory Science. *Food Rev. Int.* **2023**, *39*, 3066–3079. [CrossRef]
16. Su, K.; Sun, S.; Chen, R.Y.; Ren, M.J.; Wang, Y.B.; Bi, S.; Liu, Y. Identification of key aroma-active compounds in five types of *Zanthoxylum bungeanum* Maxim. essential oils using a molecular sensory approach. *J. Food Compos. Anal.* **2025**, *145*, 107816. [CrossRef]
17. Sgorbini, B.; Cagliero, C.; Liberto, E.; Rubiolo, P.; Bicchi, C.; Cordero, C. Strategies for Accurate Quantitation of Volatiles from Foods and Plant-Origin Materials: A Challenging Task. *J. Agric. Food Chem.* **2019**, *67*, 1619–1630. [CrossRef] [PubMed]
18. Pozo-Bayón, M.A.; Guichard, E.; Cayot, N. Flavor control in baked cereal products. *Food Rev. Int.* **2006**, *22*, 335–379. [CrossRef]
19. Xu, Y.; Bi, S.; Xiong, C.H.; Dai, Y.X.; Zhou, Q.; Liu, Y. Identification of aroma active compounds in walnut oil by monolithic material adsorption extraction of RSC18 combined with gas chromatography-olfactory-mass spectrometry. *Food Chem.* **2023**, *402*, 134303. [CrossRef] [PubMed]
20. *GB/T 12311-2012*; Sensory Analysis Method-Triangle Test. Standards Press of China: Beijing, China, 2012.
21. Conti, C.; Guarino, M.; Bacenetti, J. Measurements techniques and models to assess odor annoyance: A review. *Environ. Int.* **2020**, *134*, 105261. [CrossRef] [PubMed]
22. Shi, C.H.; Yang, F.; Yan, L.C.; Wu, J.H.; Bi, S.; Liu, Y. Characterization of key aroma-active compounds in fresh and vacuum freeze-drying mulberry by molecular sensory science methods. *J. Food Compos. Anal.* **2024**, *133*, 106387. [CrossRef]
23. Jia, Y.Y.; Wang, C.; Zhang, Y.; Deng, W.K.; Ma, Y.C.; Ma, J.F.; Han, G. The Flavor Characteristics and Metabolites of Three Commercial Dried Jujube Cultivars. *Foods* **2024**, *13*, 1193. [CrossRef] [PubMed]
24. Pino, J.A.; Quijano, C.E. Study of the volatile compounds from plum (*Prunus domestica* L. cv. Horvin) and estimation of their contribution to the fruit aroma. *Cienc. E Tecnol. Aliment.* **2012**, *32*, 76–83. [CrossRef]

https://doi.org/10.3390/separations13060177

25. Li, P.H.; Jia, Y.L.; Cai, D.L.; Wang, X.Y.; Liu, J.H.; Zhu, R.C.; Wang, Z.T.; He, Y.; Wen, L.K. Study on the relationship between flavor components and quality of ice wine during freezing and brewing of '*beibinghong*' grapes. *Food Chem.-X* **2023**, *20*, 101016. [CrossRef] [PubMed]
26. López, R.; Ezpeleta, E.; Sánchez, I.; Cacho, J.; Ferreira, V. Analysis of the aroma intensities of volatile compounds released from mild acid hydrolysates of odourless precursors extracted from *Tempranillo* and *Grenache* grapes using gas chromatography-olfactometry. *Food Chem.* **2004**, *88*, 95–103. [CrossRef]
27. Das, G.; Nath, R.; Talukdar, A.D.; Agagündüz, D.; Yilmaz, B.; Capasso, R.; Shin, H.S.; Patra, J.K. Major Bioactive Compounds from Java Plum Seeds: An Investigation of Its Extraction Procedures and Clinical Effects. *Plants* **2023**, *12*, 1214. [CrossRef] [PubMed]
28. Zheng, W.C.; Yang, X.J.; Dang, B.; Zhang, W.A.; Zhang, J.; Feng, Y.J.; Feng, Z.S.; Zheng, J. Changes in secondary metabolites of grape skins in response to different postharvest dehydration temperatures as evaluated by UPLC-Q-TOF-MS. *J. Food Meas. Charact.* **2024**, *18*, 125–136. [CrossRef]
29. Yang, P.; Song, H.L.; Wang, L.J.; Jing, H. Characterization of Key Aroma-Active Compounds in Black Garlic by Sensory-Directed Flavor Analysis. *J. Agric. Food Chem.* **2019**, *67*, 7926–7934. [CrossRef] [PubMed]
30. Sommer, S.; Hoffmann, J.L.; Fraatz, M.A.; Zorn, H. Upcycling of black currant pomace for the production of a fermented beverage with *Wolfiporia cocos*. *J. Food Sci. Technol.-Mysore* **2023**, *60*, 1313–1322. [CrossRef] [PubMed]
31. Yang, P.; Yu, M.G.; Song, H.L.; Xu, Y.Q.; Lin, Y.P.; Granvogl, M. Characterization of Key Aroma-Active Compounds in Rough and Moderate Fire Rougui Wuyi Rock Tea (*Camellia sinensis*) by Sensory-Directed Flavor Analysis and Elucidation of the Influences of Roasting on Aroma. *J. Agric. Food Chem.* **2022**, *70*, 267–278. [CrossRef] [PubMed]
32. Dudareva, N.; Klempien, A.; Muhlemann, J.K.; Kaplan, I. Biosynthesis, function and metabolic engineering of plant volatile organic compounds. *New Phytol.* **2013**, *198*, 16–32. [CrossRef] [PubMed]

https://doi.org/10.3390/separations13060177

separations

MDPI

Article

Quantitative Analysis of Eight Compounds in Traditional Korean Medicine, Gongjindan Using HPLC, UPLC–MS/MS, and GC–MS/MS Systems

Chang-Seob Seo * and Hyeun-Kyoo Shin

KM Science Research Division, Korea Institute of Oriental Medicine, Daejeon 34054, Republic of Korea
* Correspondence: csseo0914@kiom.re.kr; Tel.: +82-42-868-9361

Abstract: Gongjindan (GJD) is a traditional Korean medicine consisting of four herbal medicines and two animal-derived medicines, and is taken as a tonic in Republic of Korea. In this study, the goal was to develop and validate a simultaneous analytical method to quantify eight compounds in commercially available GJD samples using high-performance liquid chromatography (HPLC), ultra-performance liquid chromatography with tandem mass spectrometry (UPLC–MS/MS), and gas chromatography with tandem mass spectrometry (GC–MS/MS) systems. In HPLC and UPLC–MS/MS, seven components (gallic acid, 5-(hydroxymethyl)furfural, morroniside, loganin, nodakenin, decursin, and decursinol angelate) were separated and quantified using a distilled water–acetonitrile mobile phase system on a Capcell Pak UG80 C_{18} column and an Acquity UPLC BEH C18 column, respectively. Muscone was quantified using GC–MS/MS. The developed assays were validated by evaluating the linearity, limit of detection, limit of quantitation, recovery, and precision. In the regression equations of all compounds, the coefficient of determination was $\geq$0.9917, showing good linearity. The recovery was 93.70–108.17%, and the relative standard deviation values in the precision test were all <1.50%. Using the developed analysis methods, GJD samples were determined to contain the eight target compounds in concentrations from non-detected to 10.75 mg/g. The analytical assays developed and validated in this study can be used to obtain data for the quality control of commercially available GJDs and for the further expansion of efficacy and clinical studies.

Keywords: quantitative analysis; Gongjindan; HPLC; UPLC–MS/MS; GC–MS/MS

1. Introduction

Herbal medicine preparations are usually composed of two or more types of herbal medicines, animal-derived medicines, and mineral medicines, and these types of medicines have long been used to treat and prevent various diseases, and to improve health. Gongjindan (GJD) is a traditional Korean medicine consisting of four herbal medicines (*Angelica gigas* Nakai, *Cornus officinalis* Seibold & Zucc., *Panax ginseng* C.A. Meyer, and *Rehmannia glutinosa* (Gaertn.) DC.) and two animal-derived medicines (*Moschus moschifrus* L. and *Cervus nippon* Temminck), and is one of the best-known tonics in Republic of Korea [1]. When GJD has been used in animal models, liver protection [1], antioxidant [2], neuroprotective [3], anti-fatigue [4], and Alzheimer's-disease-protective [5] effects have been demonstrated.

Each of the herbal medicines that comprise GJD contain various ingredients, including coumarins (e.g., nodakenin, decursin, and decursinol angelate) from *A. gigas* [6], iridoids (e.g., loganin, morroniside, and cornuside) from *C. officinalis* [7], triterpenoid saponins (e.g., ginsenoside Rb_1 and ginsenoside Rg_1) from *P. ginseng* [8], miscellaneous compounds (e.g., 5-(hydroxymethyl)furfural; 5-HMF) from *R. glutinosa* [9], macrocyclic ketones (e.g., muscone and normuscone) and steroids (e.g., cholesterol, cholestane-3-ol, and lanosterol) from *M. moschifrus* [10–12], and peptidoglycans and gangliosides from *C. nippon* [13,14].

A range of analytical methods have been developed for the quality assessment of each constituent crude drug of GJD, based on high-performance liquid chromatography (HPLC), gas chromatography (GC), GC with mass spectrometry (GC–MS), ultra-performance liquid chromatography (UPLC), and ultra-performance liquid chromatography with tandem mass spectrometry (UPLC–MS/MS) [6–8,12,14–18]. However, the analysis of GJD as a preparation containing a combination of six types of herbal and animal-derived medicines has not yet been reported. Previously, Hong et al. [4] and Lee et al. [19] reported an analysis method using HPLC and GC–MS for GJD, which was composed of five herbal medicines, excluding *R. glutinosa*. However, *R. glutinosa*, which was excluded from these reports, is an important herbal medicine that exhibits immunomodulatory [20], hepatoprotective [21], and neuroprotective [22] activities.

In this study, a simultaneous quantitative analysis method was developed and validated for seven compounds (gallic acid, 5-HMF, morroniside, loganin, nodakenin, decursin, and decursinol angelate) using HPLC with a photodiode array detector (HPLC–PDA) and UPLC–MS/MS to evaluate the quality of GJD. In addition, muscone, a target compound of *M. moschifru*, was quantitatively analyzed and validated using gas chromatography with tandem mass spectrometry in selected ion monitoring mode (GC–MS/MS SIM).

2. Materials and Methods

2.1. Chemical and Reagents

The reference standards used in this study were purchased from standard manufacturers: gallic acid (CAS No. 149-91-7, purity of 100.0%, catalog No. G7384), 5-HMF (CAS No. 67-47-0, purity of ≥99.0%, catalog No. W501808), and cyclopentadecanone (internal standard; IS, CAS No. 502-72-7, purity of 99.2%, catalog No. C111201) from Merck KGaA (Darmstadt, Germany); morroniside (CAS No. 25406-64-8, purity of 99.9%, catalog No. BP0960), loganin (CAS No. 18524-94-2, purity of ≥98.0%, catalog No. BP0884), and decursinol angelate (CAS No. 130848-06-5, purity of 98.3%, catalog No. BP1812) from Biopurify Phytochemicals (Chengdu, China); nodakenin (CAS No. 495-31-8, purity of 99.5%, catalog No. CFN90232) from ChemFaces Biochemical (Wuhan, China); decursin (CAS No. 5928-25-6, purity of 98.7%, catalog No. DR11193) from Shanghai Sunny Biotech (Shanghai, China); muscone (CAS No. 541-91-3, purity of 97.7%, catalog No. AL-M141496) from Aladdin Biochemical Technology (Shanghai, China) (Figure S1). All other solvents (methanol, acetonitrile, and distilled water) and reagents (formic acid and acetic acid) were either HPLC- or LC–MS-grade. GJD samples were commercial products supplied by different Korean pharmaceutical companies, and the composition is shown in Table S1.

2.2. HPLC–PDA Analysis

A Shimadzu Prominence LC–20A series system (Kyoto, Japan) controlled by the LabSolution software (version 5.54, SP3) was used for the simultaneous quantitation of the seven target components (gallic acid, 5-HMF, morroniside, loganin, nodakenin, decursin, and decursinol angelate) selected from GJD samples. All analytes were separated using a Capcell Pak UG80 C_{18} analytical column (250 mm length × 4.6 mm ID, particle size 5 μm, Shiseido, Tokyo, Japan) and a mobile phase of distilled water–acetonitrile (both containing 0.1% (v/v) formic acid). Table S2 presents the HPLC operating conditions for the quantitative analysis in more detail.

For each analyte, a standard stock solution was prepared at a concentration of 1000 μg/mL using methanol. The prepared standard stock solution was kept refrigerated and serially diluted at the time of measurement. Separately, for the sample solution for HPLC quantitative analysis, 100 mg of a GJD sample was accurately taken and 10 mL of 70% methanol was added. After ultrasonic extraction at room temperature for 60 min, the sample solution was filtered using a 0.2 μm syringe filter (Pall Life Sciences, Ann Arbor, MI, USA) before HPLC analysis.

2.3. UPLC–MS/MS Analytical Conditions and Sample Preparation

The simultaneous analysis of the seven analytes in commercial GJD samples was conducted with a UPLC–MS/MS system consisting of a Waters Acquity UPLC system and a TQD MS system (Milford, MA, USA). UPLC and MS/MS analytical conditions for the simultaneous analysis are presented in Table S3, and UPLC–MS/MS MRM factors such as ion mode, MRM transition, cone voltage, and collision energy are shown in Table 1.

Table 1. Parameters for UPLC–MS/MS MRM analysis of the seven analytes.

Analyte [1]	Ion Mode	Molecular Weight	MRM Conditions	Cone Voltage (V)	Collision Energy (eV)
1	negative	170.02	169.0 → 125.0	25	15
2	negative	406.15	405.4 → 243.2	30	13
3	positive	126.03	127.0 → 109.0	20	10
4	positive	390.15	391.4 → 229.2	20	10
5	positive	408.14	409.4 → 247.2	30	15
6	positive	328.13	329.2 → 229.0	35	20
7	positive	328.13	329.2 → 229.0	35	20

[1] Gallic acid (1), morroniside (2), 5-HMF (3), loganin (4), nodakenin (5), decursin (6), and decursinol angelate (7).

The sample (approximately 50 mg) was accurately weighed into a 10 mL volumetric flask and made up with 70% methanol. Ultrasonic extraction for 5 min followed by vortexing for 1 min were carried out, and then the mixture was filtered through a 0.2 µm hydrophobic filter (SSOLKOREA Co., Ltd., Daejeon, Republic of Korea) to obtain the sample solution.

2.4. GC–MS/MS SIM Analysis

GC–MS/MS SIM analysis of muscone in GJD samples was conducted with a Shimadzu GC–MS–TQ8050 (Tokyo, Japan) consisting of a GC-2010 Plus GC system and a TQ8050 MS system. As described for the GC–MS/MS SIM analysis conditions presented in Table S4, the chromatographic separation and quantitative analysis of the IS and muscone were carried out with a Restek Rxi-5Sil MS column (20 m length, 0.18 mm ID, 0.18 µm DF, Bellefonte, PA, USA) and temperature control: 60.0 °C (1 min) → 15 °C/min → 200 °C (2 min) → 30 °C/min → 280 °C (2 min). These data were acquired and processed with the GCMSsolution software (version 4.45, Shimadzu, Tokyo, Japan).

Both stock solutions of muscone and IS were prepared at 1000 µg/mL using methanol, stored in a refrigerator, and then diluted and used during analysis while cold. Separately, for the quantification of muscone in GJD samples, approximately 50 mg of the sample was accurately weighed into a 10 mL volumetric flask, followed by ultrasonic extraction for 30 min with 50 µL of cyclopentadecanone as an IS and 3 mL of methanol. Methanol was added to adjust the volume to 10 mL to give the sample solution, which was filtered with a 0.2 µm hydrophobic filter (SSOLKOREA Co., Ltd.) before injection into the GC–MS/MS system.

2.5. Validation of Developed Analytical Method

The optimized assay was assessed by evaluating the linearity, limit of detection (LOD), limit of quantitation (LOQ), and precision based on the International Conference on Harmonisation guidelines [23]. Briefly, linearity was evaluated using the coefficient of determination (r^2) value in the regression equation for each analyte. LOD and LOQ were calculated using $3.3 \times \sigma/S$ and $10 \times \sigma/S$ (σ: the standard deviation (SD) of the y-intercept; S: the slope of the regression equation). Recovery (%) was measured by a standard addition method in which three different concentrations (low, medium, and high) of analytes were added, and it was calculated as found amount/spiked amount × 100. Finally, precision was assessed by the relative standard deviation (RSD, %) values of the analytes measured on one day and on three consecutive days (RSD (%) = SD/mean × 100). Validation of recovery and precision was performed only by HPLC and GC–MS/MS analysis methods.

3. Results and Discussion

3.1. HPLC–PDA Analysis

3.1.1. Determination of Target Compounds in GJD for Analytical Method Development Using HPLC–PDA

To identify compounds suitable for the simultaneous analysis of GJD using HPLC, the main components contained in each herbal medicine were investigated and analyzed. The compounds examined were nodakenin, decursin, and decursinol angelate from *A. gigas*; gallic acid, loganin, morroniside, and cornuside from *C. officinalis*; ginsenoside Rb_1 and ginsenoside Rg_1 from *P. ginseng*, and 5-HMF from *R. glutinosa*. As shown in Figure S2, the suitability of HPLC analysis to detect these 10 compounds in GJD samples was assessed. Except for the ginsenosides from *P. ginseng* and cornuside from *C. officinalis*, seven of the components were detected and selected as target compounds of GJD, and subsequent analysis was conducted.

3.1.2. HPLC Simultaneous Analysis Conditions

Optimal HPLC analysis conditions were set for the seven target compounds selected in Section 3.1.1. First, in column selection, C18 columns (4.6 mm × 250 mm, 5 μm) from different manufacturers, such as XBridge (Waters, Milford, MA, USA), Capcell Pak UG80 (Shiseido, Tokyo, Japan), and Quasar SPP (PerkinElmer, Buckinghamshire, UK), were compared. As shown in Figure S3, the elution and peak shape of gallic acid were more appropriately detected on the Capcell Pak UG80 column than on the other columns. Second, we sought to compare the effects of acids such as formic acid and acetic acid in the primary determined column. As a result, when formic acid was used, the capacity factor (k') of gallic acid was better than that of acetic acid, and the resolution of morroniside was better in the sample solution (Figure S4). Third, after determining the column and acid, the column oven temperature (30, 40, and 50 °C) was compared. As a result of the test, the peak of gallic acid was not sharp at 50 °C, and in the case of morroniside, it was detected overlapping with another unknown peak (Figure S5). Finally, an assay for the separation and quantification of the seven target compounds was established using a gradient elution of distilled water–acetonitrile (both containing 0.1% (v/v) formic acid) on a Capcell Pak UG80 C_{18} column (Shiseido, Tokyo, Japan) maintained at 40 °C with a Shimadzu Prominence LC-20A system coupled to a PDA (Table S2). As a result, all components were baseline separated with a resolution of ≥1.55 within 50 min, without interference from neighboring peaks (Figures 1 and S6).

3.1.3. Validation of the Developed HPLC Analytical Method

The system suitability parameters of the developed HPLC assay were calculated to be k', 0.96–15.59; separation factor (α), 1.01–1.83; theoretical plate number (N), 12,882.85–1,481,856.87; resolution (Rs), 1.55–31.28; and tailing factor (T_f), 1.01–1.24 (Table S5). The assay validation factors obtained for the simultaneous analysis of the seven target components in GJD samples are presented in Tables 2–4. For each of the calibration curves, prepared by measuring the peak areas at a series of concentration levels, the r^2 values were all 1.0000, showing excellent linearity (Table 2). The LOD and LOQ values of the components were 0.02–0.05 μg/mL and 0.06–0.16 μg/mL, respectively (Table 2). The tested recoveries (%), measured by adding three concentrations of each component, were 93.70–108.17% (RSD ≤ 2.06%, Table 3). Finally, intra- and interday precision and repeatability were evaluated for all target components and the RSD values were calculated to be ≤1.22% (Table 4 and Table S6). The satisfactory results obtained for all the validation factors confirmed that the present analytical method is appropriate for the simultaneous analysis of GJD using the seven selected target components.

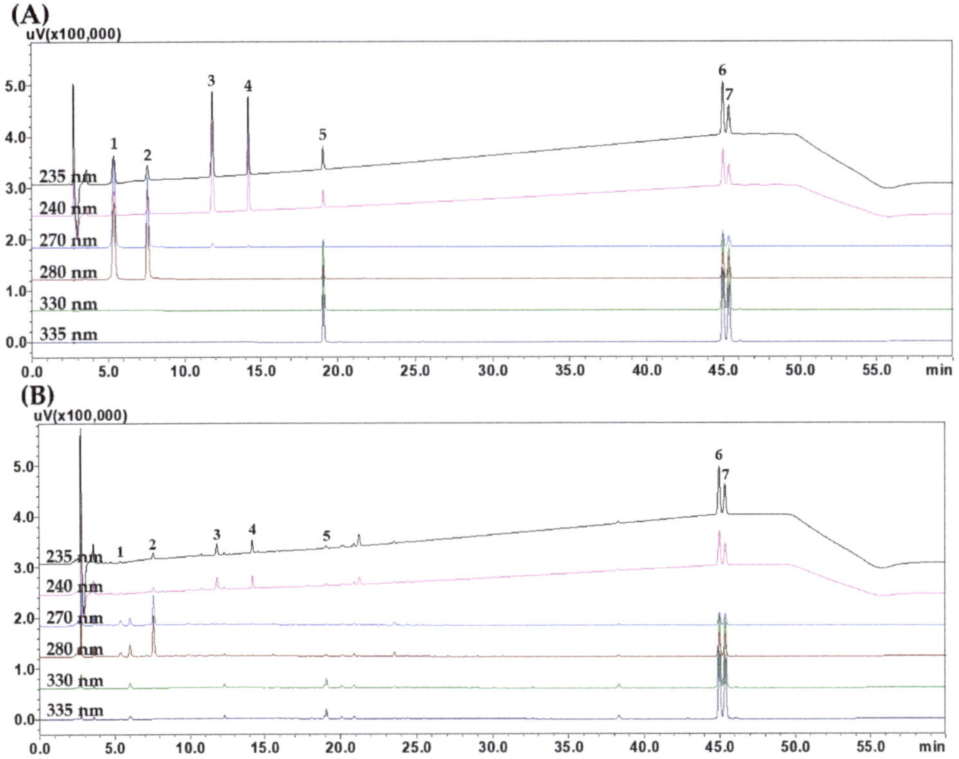

Figure 1. Representative HPLC chromatograms of the standard solution (**A**) and GJD–1 sample (**B**). Gallic acid (1), 5-(hydroxymethyl)furfural (5-HMF, 2), morroniside (3), loganin (4), nodakenin (5), decursin (6), and decursinol angelate (7). The concentrations of the compounds in the mixed standard solution were 20.00 µg/mL (compounds **2** and **5**), 30.00 µg/mL (compounds **1**, **3**, and **6**), and 40.00 µg/mL (compounds **4** and **7**).

Table 2. Method validation with respect to measured wavelength, linear range, regression equation, coefficient of determination (r^2), limit of detection (LOD), and limit of quantitation (LOQ) values for simultaneous determination of seven target components using HPLC–PDA.

Analyte [1]	Measured Wavelength (nm)	Linear Range (µg/mL)	Regression Equation [2] $y=ax+b$	r^2	LOD (µg/mL)	LOQ (µg/mL)
1	270	0.31–20.00	y = 66,489.40x + 3119.59	1.0000	0.03	0.10
2	280	0.47–30.00	y = 92,324.67x + 666.45	1.0000	0.03	0.09
3	240	0.31–20.00	y = 47,649.17x + 2644.87	1.0000	0.02	0.06
4	235	0.47–30.00	y = 22,917.18x + 1539.24	1.0000	0.04	0.13
5	335	0.31–20.00	y = 45,241.57x + 2789.69	1.0000	0.02	0.06
6	330	1.56–100.00	y = 48,629.16x + 25,391.82	1.0000	0.05	0.16
7	330	1.56–100.00	y = 32,740.22x + 12,709.59	1.0000	0.04	0.11

[1] Gallic acid (1), 5-(hydroxymethyl)furfural (5-HMF, 2), morroniside (3), loganin (4), nodakenin (5), decursin (6), and decursinol angelate (7). [2] y: peak area of compounds; x: concentration (µg/mL) of compounds.

Table 3. Recovery (%) of the seven target components for the developed HPLC method.

Analyte [1]	Spiked Amount (μg/mL)	Found Amount (μg/mL)	Recovery (%)	SD [2]	RSD (%)
1	1.00	1.04	104.00	1.69	1.62
	2.00	2.11	105.65	1.05	0.99
	4.00	4.17	104.14	0.56	0.54
2	2.00	1.94	97.19	1.74	1.79
	4.00	3.97	99.37	0.74	0.74
	8.00	7.74	96.73	0.53	0.55
3	1.00	1.03	102.81	1.38	1.34
	2.00	2.13	106.41	0.56	0.53
	4.00	4.28	107.04	0.72	0.68
4	1.00	0.97	97.16	1.75	1.80
	2.00	2.01	100.30	0.15	0.15
	4.00	4.03	100.75	1.30	1.29
5	1.00	1.02	102.07	1.62	1.59
	2.00	2.05	102.54	2.11	2.06
	4.00	4.21	105.31	1.20	1.14
6	6.00	6.03	100.29	0.83	0.83
	15.00	15.77	105.16	0.31	0.29
	30.00	32.44	108.17	0.47	0.43
7	8.00	7.50	93.70	0.62	0.67
	20.00	19.40	97.01	0.23	0.24
	40.00	40.92	102.30	0.53	0.52

[1] Gallic acid (1), 5-HMF (2), morroniside (3), loganin (4), nodakenin (5), decursin (6), and decursinol angelate (7).
[2] SD: standard deviation.

Table 4. Precision test of target compounds **1–7** in the developed HPLC method.

Analyte [1]	Conc. (μg/mL)	Intraday (n = 5)			Interday (n = 5)		
		Observed Conc. (μg/mL)	Precision (RSD, %)	Accuracy (%)	Observed Conc. (μg/mL)	Precision (RSD, %)	Accuracy (%)
1	5.00	5.19	0.17	103.88	5.21	0.53	104.27
	10.00	10.41	0.89	104.12	10.44	1.01	104.37
	20.00	20.38	0.32	101.88	20.52	0.94	102.59
2	7.50	7.89	0.28	105.13	7.91	0.96	105.52
	15.00	15.74	0.70	104.94	15.77	1.22	105.12
	30.00	30.93	0.63	103.10	31.08	1.08	103.60
3	5.00	5.18	0.17	103.62	5.21	0.52	104.17
	10.00	10.39	0.68	103.87	10.42	0.96	104.20
	20.00	20.40	0.40	101.98	20.54	0.91	102.71
4	7.50	7.78	0.18	103.80	7.81	0.48	104.11
	15.00	15.56	0.65	103.73	15.62	1.00	104.13
	30.00	30.55	0.47	101.84	30.79	0.90	102.62

Table 4. *Cont.*

Analyte [1]	Conc. (μg/mL)	Intraday (n = 5)			Interday (n = 5)		
		Observed Conc. (μg/mL)	Precision (RSD, %)	Accuracy (%)	Observed Conc. (μg/mL)	Precision (RSD, %)	Accuracy (%)
5	5.00	5.19	0.06	103.82	5.22	0.52	104.30
	10.00	10.40	0.86	104.01	10.43	1.00	104.29
	20.00	20.33	0.32	101.65	20.48	0.88	102.42
6	25.00	26.01	0.08	104.03	26.13	0.48	104.52
	50.00	52.09	0.69	104.19	52.22	0.94	104.44
	100.00	101.91	0.38	101.91	102.53	0.84	102.53
7	25.00	26.06	0.08	104.24	26.18	0.46	104.71
	50.00	52.06	0.77	104.13	52.19	0.92	104.37
	100.00	102.10	0.38	102.10	102.65	0.85	102.65

[1] Gallic acid (1), 5-HMF (2), morroniside (3), loganin (4), nodakenin (5), decursin (6), and decursinol angelate (7).

3.1.4. Simultaneous Quantitation of the Seven Target Components in GJD Samples Using the HPLC System

The HPLC assay developed herein was successfully applied to analyze the seven target components (gallic acid, 5-HMF, morroniside, loganin, nodakenin, decursin, and decursinol angelate) simultaneously in GJD samples. These analytes were monitored and quantified at the wavelengths given in Table 2, based on the UV maximum absorption wavelength. As shown in Table 5, commercially available GJD samples were found to contain 0.01–7.06 mg/g of the seven target analytes. Among these analytes, decursin and decursinol angelate, the major components of *A. gigas*, were the most abundant in all samples (2.27–7.06 mg/g). In previously reported studies [4,19], decursin derived from *A. gigas* showed high content of 7.92 mg/g. In our study, similar to the results of previous studies, decursin and decursinol angelate (isomer of decursin), the main components of *A. gigas*, were detected as the most abundant.

Table 5. Amounts (mg/g) of the seven target components in commercial products, determined using the developed HPLC method [1].

Analyte [2]	GJD 1		GJD 2		GJD 3		GJD-4		GJD-5	
	Mean (mg/g)	RSD (%)	Mean (mg/g)	RSD (%)	Mean (mg/g)	RSD (%)	Mean (mg/g)	RSD (%)	Mean (mg/g)	RSD (%)
1	0.19	0.85	0.09	0.90	0.13	0.61	0.06	0.89	0.23	0.42
2	0.77	0.29	0.15	0.23	0.01	1.66	0.08	1.69	1.99	0.16
3	0.39	0.36	0.12	0.30	0.43	0.40	0.44	0.21	0.15	0.15
4	0.58	0.41	0.42	0.27	0.71	0.29	0.65	0.03	0.38	0.19
5	0.28	0.35	0.22	0.32	0.91	0.70	0.22	0.20	0.30	0.62
6	2.78	0.18	3.24	0.10	7.06	0.11	2.76	0.18	2.57	0.03
7	4.03	0.28	3.87	0.21	6.54	0.03	2.35	0.20	3.24	0.06

[1] GJD-1 to GJD-5: commercial products supplied by Korean pharmaceutical companies. [2] Gallic acid (1), 5-HMF (2), morroniside (3), loganin (4), nodakenin (5), decursin (6), and decursinol angelate (7).

3.2. UPLC–MS/MS Analysis

3.2.1. UPLC–MS/MS Multiple Reaction Monitoring (MRM) Method for Simultaneous Analysis

UPLC–MS/MS quantitative analysis was performed on the seven target components selected in the HPLC analytical method. A UPLC–MS/MS system (Waters, Milford, MA, USA) consisting of a Waters Acquity UPLC I–Class system and a tandem quadrupole MS detector system with an electrospray ionization source was used. UPLC–MS/MS MRM analysis was conducted with a Waters Acquity UPLC BEH C_{18} column (2.1 mm × 100 mm, 1.7 μm) and a distilled water–acetonitrile mobile-phase system (Tables S3 and 1). Gallic acid and morroniside were detected in negative ion mode as $[M-H]^-$, and 5-HMF, loganin, nodakenin, decursin, and decursinol angelate were detected in positive ion mode as $[M+H]^+$, within 10.0 min (Figures 2 and S7).

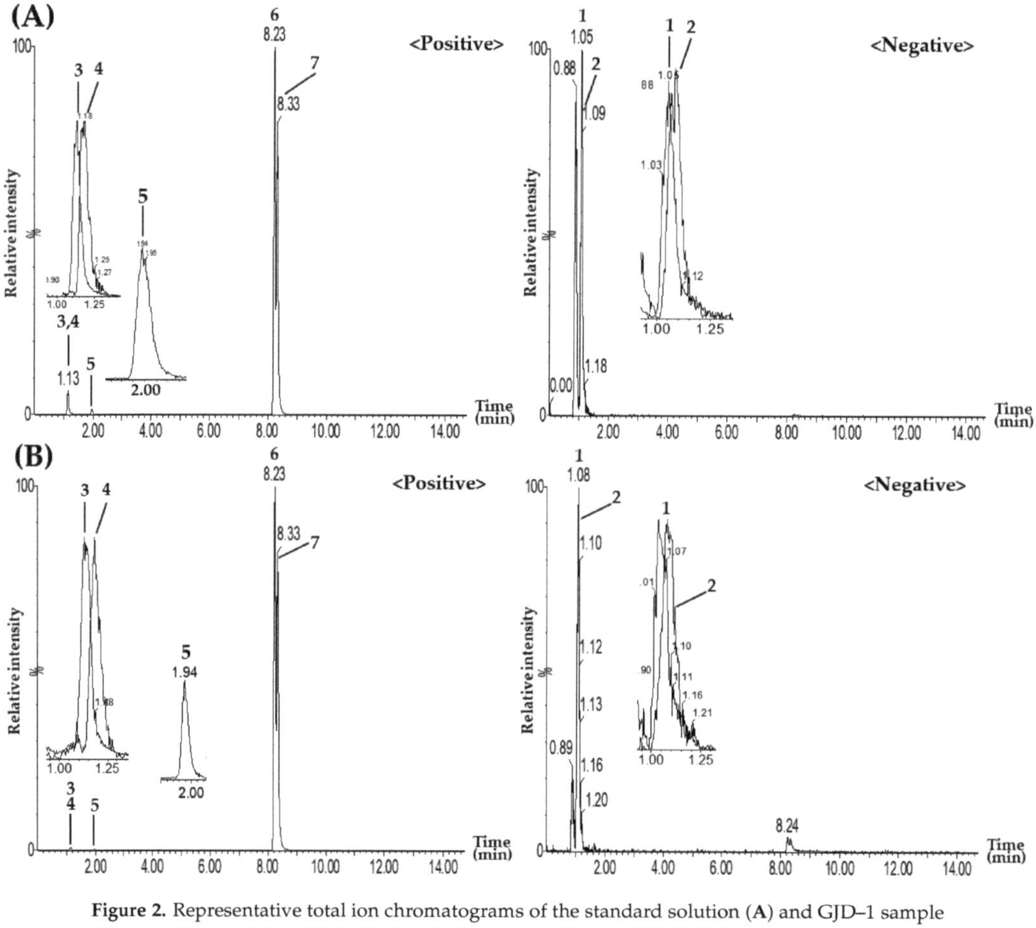

Figure 2. Representative total ion chromatograms of the standard solution (**A**) and GJD–1 sample (**B**) using UPLC–MS/MS MRM. Gallic acid (1), morroniside (2), 5-HMF (3), loganin (4), nodakenin (5), decursin (6), and decursinol angelate (7). The concentration of all analytes in the mixed standard solution was 5.00 μg/mL.

For quantitative analysis, the MRM conditions (precursor ion (Q1) and product ion (Q3)) of the seven analytes were set as shown in Table 5. The Q3 peaks of gallic acid and 5-HMF were set to m/z 125.0 and 109.0 in the form of $[M\text{–}H\text{–}COO^-]^-$ and $[M\text{+}H\text{–}H_2O]^+$, respectively, in which a carboxyl group (COO^-) and water (H_2O) molecule were lost from each Q1 peak [24,25]. For MRM transitions of iridoid-type compounds morroniside and loganin, the Q1 peak was detected at m/z 405.4 and 391.4 in the form of $[M–H]^-$ and $[M+H]^+$, respectively. All of the detected ions were in the form of aglycone in which the glucose group was lost, and Q3 peaks were detected at m/z 243.2 ($[M–H–Glu]^-$) and 229.2 ($[M+H–Glu]^+$) [26,27]. For nodakenin, the Q3 peak was set at m/z 247.2, corresponding to the $[M+H-Glu]^+$ form, where one molecule of glucose was removed from the Q1 peak at m/z 409.4 [28]. The Q3 peak for the two coumarin derivatives, decursin and decursinol angelate, was set at m/z 229.0 for the $[M+H-C_5H_8O_2]^+$ form, with the isoprenyl backbone and hydroxy group removed from the Q1 peak at m/z 329.2 [15].

3.2.2. Quantitation of the Seven Target Components in GJD Samples

The values for the regression equations r^2, LOD, LOQ, and retention times of the seven analytes prepared at a concentration of 0.10–5.00 μg/mL are presented in Table 6. In GJD samples, concentrations of the seven investigated analytes ranged from non-detected to 10.75 mg/g (Table 7). As a result of UPLC–MS/MS analysis, decursin and decursinol angelate, the main components of *A. gigas*, were found with the highest content (3.86–10.75 mg/g and 2.95–7.81 mg/g, respectively), which is consistent with the results obtained by HPLC analysis.

Table 6. Retention time, linear range, regression equation, r^2, LOD, and LOQ values for simultaneous determination of the seven analytes using the UPLC–MS/MS method.

Analyte [1]	Retention Time (min)	Linear Range (μg/mL)	Regression Equation [2] $y = ax + b$	r^2	LOD (ng/mL)	LOQ (ng/mL)
1	1.06	0.10–5.00	$y = 0.61x - 14.81$	0.9967	3.27	9.82
2	1.08	0.10–5.00	$y = 0.47x - 6.36$	0.9917	3.50	10.49
3	1.13	0.10–5.00	$y = 23.77x + 425.4$	0.9929	1.61	4.82
4	1.18	0.10–5.00	$y = 1.01x + 12.56$	0.9921	9.92	29.75
5	1.94	0.10–5.00	$y = 5.33x + 114.23$	0.9943	3.56	10.68
6	8.23	0.10–5.00	$y = 189.03x + 1684.86$	0.9945	0.15	0.46
7	8.33	0.10–5.00	$y = 168.94x + 2700.37$	0.9959	0.16	0.48

[1] Gallic acid (1), morroniside (2), 5-HMF (3), loganin (4), nodakenin (5), decursin (6), and decursinol angelate (7).
[2] y: peak area of compounds; x: concentration (μg/mL) of compounds.

Table 7. Amounts (mg/g) of the seven analytes in GJD–1 to GJD–5 samples determined using the UPLC–MS/MS method [1].

Analyte [2]	GJD–1		GJD–2		GJD–3		GJD–4		GJD–5	
	Mean (mg/g)	RSD (%)	Mean (mg/g)	RSD (%)	Mean (mg/g)	RSD (%)	Mean (mg/g)	RSD (%)	Mean (mg/g)	RSD (%)
1	0.30	3.97	0.11	2.31	0.19	1.16	0.06	1.57	0.37	2.62
2	0.67	2.45	0.29	1.84	0.73	0.69	0.78	4.79	0.24	3.79
3	0.44	0.93	0.11	4.57	ND [3]	–	0.02	3.58	1.34	0.85
4	0.64	3.20	0.57	2.41	0.91	3.44	0.78	1.95	0.46	2.62
5	0.75	2.47	0.64	1.30	2.41	4.65	0.77	1.71	0.87	2.80
6	4.03	0.67	4.89	2.23	10.75	1.20	4.44	1.53	3.86	0.81
7	4.49	0.49	4.45	3.08	7.81	2.16	2.95	0.24	3.62	0.33

[1] GJD–1 to GJD–5: commercial products supplied by Korean pharmaceutical companies. [2] Gallic acid (1), 5-HMF (2), morroniside (3), loganin (4), nodakenin (5), decursin (6), and decursinol angelate (7). [3] ND: not detected.

3.3. GC–MS/MS Analysis

3.3.1. Establishment of the GC–MS/MS Method

Muscone, which was selected as a target compound for *M. moschifrus* in GJD samples, was quantitatively analyzed with cyclopentadecanone as IS using the GC–MS/MS SIM method. For GC–MS/MS SIM quantitation, the quantitative ions of muscone and IS were set to m/z 238.23 → 85 and m/z 224.21 → 55, respectively, and analyzed under the conditions shown in Table S4.

3.3.2. GC–MS/MS Method Validation for Analysis of Muscone

Data for calibration curves, LOD, LOQ, recovery, and precision were recorded to validate the method. The regression equation ($y = 1.02x - 1.62 \times 10^{-2}$) of the calibration curve was prepared using the area ratio of IS and muscone in the range of 0.5–1.0 μg/mL, and the r^2 value was 0.9998, showing good linearity. LOD and LOQ values were calculated with signal-to-noise ratios of 3 and 10, respectively, and were 0.13 μg/mL and 0.38 μg/mL, respectively. Recovery (%) was found to be 99.13–107.33%, and the RSDs for intra- and interday precision were all ≤1.44% (Tables 8 and 9). The above data confirm that the assay established for the quantitation of muscone in GJD samples was appropriate.

Table 8. Recovery (%) of muscone by the GC–MS/MS method (n = 5).

Analyte	Spiked Amount (μg/mL)	Found Amount (μg/mL)	Recovery (%)	SD	RSD (%)
Muscone	110.00	109.00	99.13	2.53	2.55
	275.00	277.30	100.84	1.49	1.48
	550.00	590.30	107.33	0.83	0.77

Table 9. Verification of intra- and interday precision for muscone using the GC–MS/MS method (n = 5).

Analyte	Conc. (μg/mL)	Intraday (n = 5)			Interday (n = 5)		
		Observed Conc. (μg/mL)	Precision (RSD, %)	Accuracy (%)	Observed Conc. (μg/mL)	Precision (RSD, %)	Accuracy (%)
Muscone	2.50	2.49	0.67	99.72	2.52	1.30	100.68
	5.00	4.91	0.97	98.19	4.97	0.95	99.40
	10.00	10.26	0.66	102.56	10.22	1.44	102.19

3.3.3. Quantitation of Muscone in GJD Samples Using the GC–MS/MS Method

The established GC–MS/MS SIM method was applied to the efficient analysis of real GJD samples, and both the IS and muscone were detected at 10.02 min and 10.26 min without interference from other peaks, respectively (Figures 3 and S8). Muscone was detected at 0.44–0.52 mg/g in samples of GJD–1 to GJD–5 (Table 10).

Table 10. Amounts (mg/g) of muscone in commercial products determined using the GC–MS/MS method [1].

Analyte	GJD–1		GJD–2		GJD–3		GJD–4		GJD–5	
	Mean (mg/g)	RSD (%)	Mean (mg/g)	RSD (%)	Mean (mg/g)	RSD (%)	Mean (mg/g)	RSD (%)	Mean (mg/g)	RSD (%)
Muscone	0.48	3.43	0.51	14.31	0.52	4.13	0.49	4.61	0.44	2.45

[1] GJD–1 to GJD–5: commercial products supplied by Korean pharmaceutical companies.

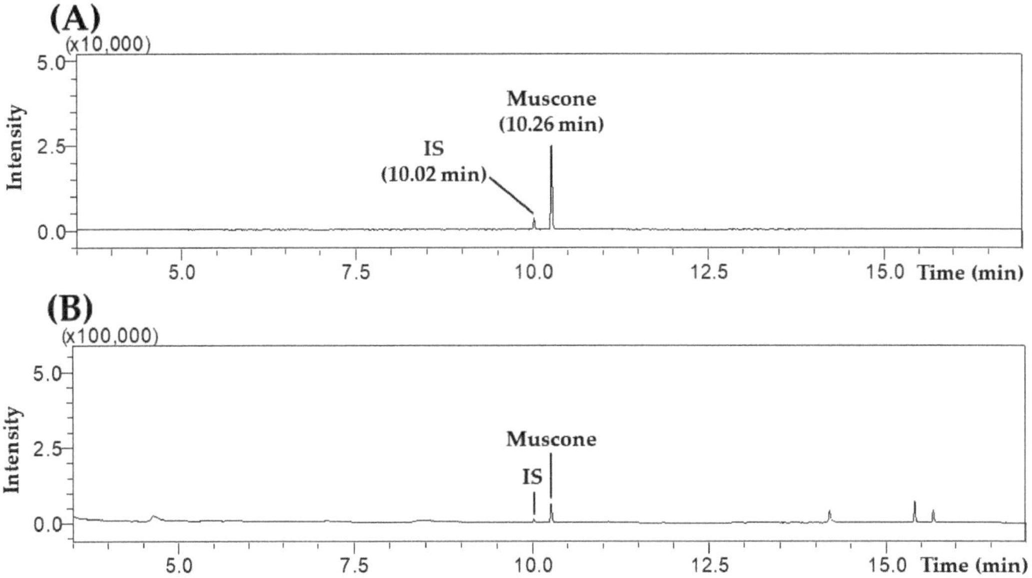

Figure 3. Representative GC–MS/MS chromatograms of the standard solution (**A**) and GJD–1 sample (**B**). The concentrations of muscone and IS in the mixed standard solution were 5.00 μg/mL and 1.00 μg/mL, respectively.

4. Conclusions

In this study, a method has been developed and validated for the simultaneous analysis of eight components in commercially available GJD, a traditional herbal prescription widely used as a tonic, using HPLC, UPLC–MS/MS, or GC–MS/MS systems. These analytical methods were appropriately applied to effectively analyze GJD samples. Among the seven analytes, except for muscone, in these analysis methods, decursin and decursinol angelate, the main components of A. gigas, were detected the most abundantly by the HPLC and UPLC–MS/MS methods. In general, the UPLC–MS/MS MRM method has the advantages of a shorter analysis time, higher sensitivity, and easy multi-component analysis compared to the HPLC method. Nevertheless, as an analysis method for the quality control of GJD samples using the seven analytes, it is considered that the analytical method using HPLC, which is the most widely used and easy to operate today, is more appropriate. Finally, these assays can be used to obtain basic data for clinical trials and for efficacy verification.

Supplementary Materials: The following supporting information can be downloaded at: https://www.mdpi.com/article/10.3390/separations10040231/s1, Figure S1: Chemical structures of the eight target components and internal standard (IS) in GJD; Figure S2: HPLC chromatograms of the solution of standard mixture (**A**) and 70% methanol solution of GJD–1 sample (**B**). Gallic acid (1), 5-(hydroxymethyl)furfural (2), morroniside (3), loganin (4), nodakenin (5), cornuside (6), ginsenoside Rg_1 (7), ginsenoside Rb_1 (8), decursin (9), and decursinol angelate (10); Figure S3: HPLC chromatograms according to columns. Standard mixture (**A**) and GJD sample (**B**) using Capcell Pak UG80 C_{18} column, standard mixture (**C**) and GJD sample (**D**) using Quasar SPP C_{18} column, and standard mixture (**C**) and GJD sample (**D**) using XBridge C_{18} column; Figure S4: HPLC chromatograms according to acids. Standard mixture (**A**) and GJD sample (**B**) using 0.1% (*v*/*v*) formic acid and standard mixture (**C**) and GJD sample (**D**) using 1.0% (*v*/*v*) acetic acid; Figure S5: HPLC chromatograms according to column temperatures. Standard mixture (**A**) and GJD sample (**B**) at 30 °C, standard mixture (**C**) and GJD sample (**D**) at 40 °C, and standard mixture (**E**) and GJD sample (**F**) at 50 °C; Figure S6: HPLC chromatograms of GJD–2 to GJD–5 samples (**A**) and each reference compound (**B**). Gallic acid (1), 5-(hydroxymethyl)furfural (2), morroniside (3), loganin (4), nodakenin (5), decursin (6), and decursinol angelate (7); Figure S7: Total ion chromatograms of GJD–2 to GJD–5 samples by UPLC–MS/MS MRM method. Gallic acid (1), morroniside (2), 5-(hydroxymethyl)furfural (3), loganin (4), nodakenin (5), decursin (6), and decursinol angelate (7); Figure S8: GC–MS/MS chromatograms of GJD–2 to GJD–5 samples; Table S1: Composition of Gongjindan one-pill; Table S2: Analytical conditions for simultaneous determination of the seven components in GJD by HPLC; Table S3: LC–MS/MS MRM analysis conditions for quantification of the target components in GJD; Table S4: GC–MS/MS selective ion monitoring (SIM) analytical conditions for quantification of muscone in GJD samples; Table S5: System suitability for HPLC simultaneous analysis of the seven components using a standard solution; Table S6: Repeatability of retention time and peak area of the seven analytes using HPLC (n = 6).

Author Contributions: Conceptualization, C.-S.S. and H.-K.S.; methodology, validation, formal analysis, investigation, data curation, writing—original draft preparation, writing—review and editing, C.-S.S.; project administration, funding acquisition, H.-K.S. All authors have read and agreed to the published version of the manuscript.

Funding: This research was funded by the Korea Institute of Oriental Medicine (KSN2022310 and KSN1823311).

Institutional Review Board Statement: Not applicable.

Informed Consent Statement: Not applicable.

Data Availability Statement: All data of this study can be found in this paper.

Conflicts of Interest: The authors declare no conflict of interest.

Sample Availability: Samples of the compounds are available after consultation with the authors.

References

1. Han, M.G.; Kim, K.S.; Joo, J.H.; Choi, H.S.; Kim, S.M. Favorable hepatoprotective effects of Gongjin-dan on the acute ethanol—induced liver damaged C57BL/6 mice. *J. Physiol. Pathol. Korean Med.* **2016**, *30*, 279–288. [CrossRef]
2. Choi, K.H.; Park, C.S. An analysis of the Gongjindan's ingredients and its efficacy on anti-oxidation. *Kor. J. Herbol.* **2007**, *22*, 51–63.
3. Moon, E.; Her, Y.; Lee, J.B.; Park, J.H.; Lee, E.H.; Kim, S.H.; Oh, M.S.; Jang, C.G.; Kim, S.Y. The multi-herbal medicine Gongjin-dan enhances memory and learning tasks via NGF regulation. *Neurosci. Lett.* **2009**, *466*, 114–119. [CrossRef] [PubMed]
4. Hong, S.S.; Lee, J.; Lee, J.S.; Lee, H.W.; Kim, H.G.; Lee, S.K.; Park, B.K.; Son, C.G. The traditional drug *Gongjin-dan* ameliorates chronic fatigue in a forced-stress mouse exercise model. *J. Ethnopharmacol.* **2015**, *168*, 268–278. [CrossRef]
5. Hwang, S.M.; Chung, D.K. The effects of Kongjin-dan (KJD) on the Alzheimer's disease model induced by CT105. *J. Orient. Neuropsychiatry* **2004**, *15*, 103–118.
6. Jeong, S.Y.; Kim, H.M.; Lee, K.H.; Kim, K.Y.; Huang, D.S.; Kim, J.H.; Seong, R.S. Quantitative analysis of marker compounds in *Angelica gigas*, *Angelica sinensis*, and *Angelica acutiloba* by HPLC/DAD. *Chem. Pharm. Bull.* **2015**, *63*, 504–511. [CrossRef]
7. Jiang, Y.; Chen, H.; Wang, L.; Zou, J.; Zheng, X.; Liu, Z. Quality evaluation of polar and active components in crude and processed *Fructus corni* by quantitative analysis of multicomponents with single marker. *J. Anal. Methods Chem.* **2016**, *2016*, 6496840. [CrossRef]
8. Shan, S.M.; Luo, J.G.; Huang, F.; Kong, L.Y. Chemical characteristics combined with bioactivity for comprehensive evaluation of *Panax ginseng* C.A. Meyer in different ages and seasons based on HPLC-DAD and chemometric methods. *J. Pharm. Biomed. Anal.* **2014**, *89*, 76–82. [CrossRef]
9. Lee, J.Y.; Lee, E.J.; Kim, J.S.; Lee, J.H.; Kang, S.S. Phytochemical studies on Rehmanniae Radix Preparata. *Kor. J. Pharmacogn.* **2011**, *42*, 117–126.
10. Fujimoto, S.; Yoshikawa, K.; Itoh, M.; Kitahara, T. Synthesis of (*R*)- and (*S*)-muscone. *Biosci. Biotechnol. Biochem.* **2002**, *66*, 1389–1392. [CrossRef]
11. Oh, S.H. *The Korean Herbal Pharmacopoeia*, 5th ed.; Shinilbooks: Seoul, Republic of Korea, 2016; pp. 97–99.
12. Liu, K.; Xie, L.; Deng, M.; Zhang, X.; Luo, J.; Li, X. Zoology, chemical composition, pharmacology, quality control and future perspective of Musk (Moschus): A review. *Chin. Med.* **2021**, *16*, 46. [CrossRef] [PubMed]
13. Jiang, W.; Yin, Y.; Zhou, Y.; He, G.; Qi, Y. Isolation and characterization of peptidoglycan recognition protein 1 from antler base of silka deer (*Cervus nippon*). *Int. J. Biol. Macromol.* **2014**, *64*, 313–318. [CrossRef] [PubMed]
14. Choi, H.O.; Kim, J.W.; Jo, S.J.; Kim, J.H.; Han, B.H. A new HPLC-analytical method for total sphingosine contents as an indirect index for the ganglioside contents of deer antlers. *Nat. Prod. Sci.* **2011**, *17*, 315–320.
15. Ahn, M.J.; Lee, M.K.; Kim, Y.C.; Sung, S.H. The simultaneous determination of coumarins in *Angelica gigas* root by high performance liquid chromatography–diode array detector coupled with electrospray ionization/mass spectrometry. *J. Pharm. Biomed. Anal.* **2008**, *46*, 258–266. [CrossRef] [PubMed]
16. Hwang, S.Y.; Hwang, H.Y.; Choi, W.H.; Jung, H.J.; Huh, J.D.; Lee, K.S.; Ro, J.S. Quantitative determination of 5-hydroxymethyl-2-furaldehyde in the Rehmanniae Radix Preparata samples at various processing stages. *Kor. J. Pharmacogn.* **2001**, *32*, 116–120.

17. Jin, C.; Yan, C.; Luo, Y.; Li, B.; He, J.; Xiao, X. Fast and direct quantification of underivatized muscone by ultra performance liquid chromatography coupled with evaporative light scattering detection. *J. Sep. Sci.* **2013**, *36*, 1762–1767. [CrossRef] [PubMed]
18. Zhang, H.; Tao, Y.; Hong, X.; Wang, Z. Steroids in musk by gas chromatography/mass spectrometry. *Chin. Tradit. Pat. Med.* **2005**, *27*, 79–83.
19. Lee, J.S.; Hong, S.S.; Kim, H.G.; Lee, H.W.; Kim, W.Y.; Lee, S.K.; Son, C.G. *Gongjin-dan* enhances hippocampal memory in a mouse model of scopolamine-induced amnesia. *PLoS ONE* **2016**, *11*, e0159823. [CrossRef]
20. Huang, Y.; Jiang, C.; Hu, Y.; Zhao, X.; Shi, C.; Yu, Y.; Liu, C.; Tao, Y.; Pan, H.; Feng, Y.; et al. Immunoenhancement effect of *Rehmannia glutinosa* polysaccharide on lymphocyte proliferation and dendritic cell. *Carbohydr. Polym* **2013**, *96*, 516–521. [CrossRef]
21. Liu, Y.F.; Liang, D.; Luo, H.; Hao, Z.Y.; Wang, Y.; Zhang, C.L.; Zhang, Q.J.; Chen, R.Y.; Yu, D.Q. Hepatoprotective iridoid glycosides from the roots of *Rehmannia glutinosa*. *J. Nat. Prod* **2012**, *75*, 1625–1631. [CrossRef]
22. Zhang, X.; Zhang, A.; Jiang, B.; Bao, Y.; Wang, J.; An, L. Further pharmacological evidence of the neuroprotective effect of catalpol from *Rehmannia glutinosa*. *Phytomedicine* **2008**, *15*, 484–490. [CrossRef] [PubMed]
23. International Conference on Harmonisation (ICH). *Guidance for Industry, Q2B, Validation of Analytical Procedures: Methodology*; Food and Drug Administration: Rockville, MD, USA, 1996.
24. Jiang, M.; Zhou, M.; Han, Y.; Xing, L.; Zhao, H.; Dong, L.; Bai, G.; Luo, G. Identification of NF-κB inhibitors in Xuebijing injection for sepsis treatment based on bioactivity-integrated UPLC-Q/TOF. *J. Ethnopharmacol.* **2013**, *147*, 426–433. [CrossRef] [PubMed]
25. Ye, J.; Zhang, X.; Dai, W.; Yan, S.; Huang, H.; Liang, X.; Li, Y.; Zhang, W. Chemical fingerprinting of Liuwei Dihuang Pill and simultaneous determination of its major bioactive constituents by HPLC coupled with multiple detections of DAD, ELSD and ESI-MS. *J. Pharm. Biomed. Anal.* **2009**, *49*, 638–645. [CrossRef] [PubMed]
26. Dan, G.; Cho, C.W.; Vinh, L.B.; Kim, J.H.; Cho, K.W.; Kim, Y.H.; Kang, J.S. Simultaneous qualitative and quantitative analysis of morroniside and hederacoside D in extract mixture of *Cornus officinalis* and *Stauntonia hexaphylla* leaves to improve benign prostatic hyperplasia by HPLC–UV. *Anal. Sci. Technol.* **2020**, *33*, 224–231.
27. Song, Y.; Li, S.L.; Wu, M.H.; Li, H.J.; Li, P. Qualitative and quantitative analysis of iridoid glycosides in the flower buds of *Lonicera* species by capillary high performance liquid chromatography coupled with mass spectrometric detector. *Anal. Chim. Acta* **2006**, *564*, 211–218. [CrossRef]
28. Kim, M.K.; Yang, D.H.; Jung, M.; Jung, E.H.; Eom, H.Y.; Suh, J.H.; Min, J.W.; Kim, U.; Min, H.; Kim, J.; et al. Simultaneous determination of chromones and coumarins in Radix Saposhnikoviae by high performance liquid chromatography with diode array and tandem mass detectors. *J. Chromatogr. A* **2011**, *1218*, 6319–6330.

separations

MDPI

Review

Chromatographic Applications Supporting ISO 22002-100:2025 Requirements on Allergen Management, Food Fraud, and Control of Chemical and Packaging-Related Contaminants

Eftychia G. Karageorgou [1], Nikoleta Andriana F. Ntereka [2] and Victoria F. Samanidou [3,*]

[1] NBIS P.C., Cargo Inspections and ISO Certification, GR-54351 Thessaloniki, Greece; ekarageorgou@nbis.gr
[2] Department of Food Science and Technology, International Hellenic University, GR-57400 Thessaloniki, Greece; nintereca@ihu.gr
[3] Laboratory of Analytical Chemistry, School of Chemistry, Aristotle University of Thessaloniki, GR-54124 Thessaloniki, Greece
* Correspondence: samanidu@chem.auth.gr

Abstract

ISO 22002-100:2025 introduces stringent and more technically explicit prerequisite programme (PRP) requirements for allergen management, food fraud mitigation, and the control of chemical and packaging-related contaminants across the food, feed, and packaging supply chain. This review examines how advanced chromatographic methods provide the analytical basis required to meet these requirements and to support alignment with GFSI-recognized certification schemes. Recent applications of liquid and gas chromatography coupled with mass spectrometry for allergen quantification, authenticity assessment, and the determination of packaging migrants, auxiliary chemical residues, lubricants, and indoor pest-control pesticides are presented to demonstrate their relevance as verification tools. Across these PRP-related controls, chromatographic methods enable trace-level detection, structural specificity, and reproducible measurement performance, thereby shifting PRP compliance from a documentation-based activity to a process verified through measurable analytical evidence. The review highlights significant progress in method development and simultaneous multi-target analytical approaches while also identifying remaining challenges related to matrix-appropriate validation, harmonization, and analytical coverage for chemical contamination, which is now formally defined as a measurable PRP requirement under ISO 22002-100:2025. Overall, the findings demonstrate that chromatographic analysis has become essential to demonstrating PRP effectiveness under ISO 22002-100:2025, supporting the broader shift toward evidence-based, scientifically robust food safety assurance.

Keywords: ISO 22002-100:2025; GFSI; prerequisite programmes; chromatographic verification; allergens; food fraud; chemical contaminants; packaging migrants; LC-MS/MS; GC-MS/MS

1. Introduction

The prevention of contamination, adulteration, and allergen cross-contact has long been a fundamental objective of food safety management systems. Beyond microbiological hazards, chemical and physical hazards, ranging from undeclared allergens and fraudulent substitutions to packaging-derived contaminants and process-related chemical residues, including cleaning agents, disinfectants, lubricants, and pest control substances, require continuous analytical verification to ensure product integrity and consumer protection. To address these multifactorial risks within the global food supply chain, the ISO 22000 family provides a harmonized management system integrating risk assessment with principles of continual improvement, while the ISO 22002 series specifies prerequisite programmes that translate these management principles into operational and environmental controls. These PRPs define the operational and environmental prerequisites needed to control hazards that extend beyond the scope of Critical Control Points [1].

ISO 22002-100:2025 (Food, Feed, and Packaging Supply Chain) represents a major step toward an integrated approach to risk management across all organizations in the agri-food chain [1]. It defines common prerequisite requirements applicable to the food, feed, and packaging supply chain and forms the basis for all sector-specific standards in the series. Alongside ISO 22002-100:2025, the recently issued ISO 22002-7:2025 for retail and wholesale activities further strengthens the consistency of requirements throughout the agri-food supply chain [2]. Together, these standards consolidate and update previous sector-specific requirements related to food manufacturing hygiene, packaging processes, and retail or distribution operations, creating a unified reference for prerequisite programmes across the food and packaging sectors. The transition from the previous Technical Specifications (ISO/TS 22002 series) to a full International Standard strengthens the authority and global recognition of these prerequisite programme requirements.

Focusing on ISO 22002-100:2025, the standard expands the preventive scope to three technically demanding areas:

(a) allergen management, requiring validated analytical verification of cleaning and segregation procedures,
(b) food fraud mitigation and authenticity verification, demanding vulnerability assessments supported by appropriate analytical verification methods; and
(c) chemical contamination and packaging migration control, including non-intentionally added substances (NIASs), residues of cleaning and disinfection agents, technical fluids, lubricants, and pest control chemicals [1].

In addition to allergen management, food fraud prevention, chemical contamination, and packaging safety, the revised standard strengthens requirements on food defence, waste management, and equipment suitability, reflecting a more integrated approach to food safety risks. ISO 22002-100:2025 adopts the Annex SL structure for management systems, ensuring consistency in terminology, clause alignment, and risk-based methodology with ISO 22000:2018 and ISO 22003-1:2022 [3,4]. The standard emphasizes a life-cycle approach to hazard prevention, integrating supplier assurance, analytical verification, and traceability into a unified compliance model. It also introduces explicit requirements for analytical verification, recommending the use of standardized and sensitive instrumental techniques for detecting allergens, authenticity markers, and chemical hazards, including chromatographic determination of residues arising from sanitation chemicals, technical substances, and packaging-related migrants [1]. Such provisions establish analytical chemistry, particularly chromatographic science, from a supporting activity to a key element in verifying PRP effectiveness.

Recent research linking regulatory compliance with ISO-based food safety management systems [5] shows that the development of the ISO 22000 and 22002 series reflects a broader shift toward measurable, evidence-based control. Analytical data derived from chromatographic and spectrometric methods are increasingly used as objective indicators connecting operational performance with product safety outcomes. This integration of analytical verification into risk management reinforces the expectation that control measures must be scientifically justified and reproducible.

https://doi.org/10.3390/separations13010039

Therefore, chromatographic methods such as HPLC and GC, along with complementary detection techniques including spectroscopic and hybrid analytical systems, provide a fundamental analytical basis for implementing and verifying the new ISO requirements. They enable trace-level detection of allergens, identification of authenticity markers, and monitoring of chemical contaminants as well as packaging migrants, transforming qualitative management principles into measurable compliance evidence. The following sections examine recent chromatographic advances and their role in supporting the implementation of ISO 22002-100:2025, from analytical verification of allergens and authenticity to the chromatographic control of packaging-derived migrants, process chemicals, and pest control residues, as well as broader aspects of system integration and compliance alignment with Global Food Safety Initiative (GFSI) requirements.

Accordingly, the scope of this review is to present recent compliance-relevant requirements introduced or reinforced by ISO 22002-100:2025 in relation to PRP implementation and verification within food safety management, and to discuss how these expectations align with GFSI-recognized schemes. Within this context, emphasis is placed on where chromatographic evidence can provide the accuracy, specificity, and reproducibility required in industrial practice to verify PRP effectiveness (e.g., allergen management, authenticity-related verification, and control of chemical and packaging-related contaminants), rather than providing an overview of chromatographic technologies used for these determinations. More specifically, contemporary chromatographic methods applied in these verification activities are discussed in this review. Finally, areas where ISO 22002-100:2025 verification expectations may not yet be fully supported by matrix-appropriate chromatographic methods are identified, thereby highlighting analytical gaps and priorities for further method development.

2. Chromatographic Methods Supporting Allergen Management

2.1. Regulatory and Traceability Context for Allergen Control

Allergen control in food production is fundamentally guided by Regulation (EU) No 1169/2011 on the provision of food information to consumers, which mandates the clear labelling of substances known to cause allergies or intolerances. Annex II of the Regulation lists fourteen major allergens, namely, cereals containing gluten, eggs, milk, soybeans, peanuts, tree nuts, fish, shellfish, celery, mustard, sesame, lupin, sulphur dioxide/sulphites, and molluscs, that must always be declared regardless of their concentration in the product [6]. These legislative requirements establish the basis for preventive allergen management throughout the food supply chain. In alignment, ISO 22002-100:2025 requires organizations to implement validated cleaning, segregation, and traceability measures within their PRPs to prevent cross-contact and ensure compliance with applicable food information legislation [1]. The integration of regulatory and analytical verification measures under this standard reinforces evidence-based allergen control as a core component of modern food safety management systems. The fourteen allergens listed in Annex II of Regulation (EU) No 1169/2011 are summarized in Table 1, together with their main food sources and categories.

https://doi.org/10.3390/separations13010039

Table 1. Overview of the fourteen major allergens listed in Annex II of Regulation (EU) No 1169/2011.

Category	Allergen	Typical Food Sources
Cereals containing gluten	Wheat, rye, barley, oats	Bread, pasta, beer, breakfast cereals, malt-based products, sauces
Crustaceans	Shrimp, crab, lobster	Seafood dishes, sauces, soups
Eggs	Egg proteins (ovalbumin, ovomucoid)	Bakery products, mayonnaise, sauces, desserts
Fish	Cod, salmon, tuna	Processed fish products, ready meals, sauces
Peanuts	Groundnuts (peanuts)	Peanut butter, snacks, confectionery
Soybeans	Soy proteins (including lecithins)	Tofu, soy-based sauces, meat substitutes, processed foods
Milk	Casein, β-lactoglobulin	Milk, cheese, butter, yogurt, dairy-based desserts
Nuts	Tree nuts (almond, hazelnut, walnut, cashew, pistachio, Brazil nut, macadamia)	Confectionery, desserts, bakery products
Celery	Celery (root and stalk parts)	Soups, sauces, spice mixes, ready meals
Mustard	Mustard seeds	Condiments, dressings, processed foods
Sesame seeds	Sesame seeds and derived products (e.g., tahini, oil)	Bakery products, spreads, sauces
Sulphur dioxide/sulphites	Sulphur dioxide and sulphite preservatives (>10 mg/kg or L)	Dried fruits, wine, beer, processed vegetables
Lupin	Lupin flour and derived ingredients	Gluten-free products, bakery goods, pasta substitutes
Molluscs	Wheat, rye, barley, oats	Seafood dishes, mixed seafood products

Effective allergen management relies not only on regulatory compliance but also on the traceability and quantification of allergenic ingredients throughout the food chain. Traceability provides the means to document ingredient origins, processing steps, and potential cross-contact points, thereby ensuring transparency and accountability within allergen control systems [7]. Modern approaches integrate traceability data with risk-based management tools to establish measurable thresholds that define acceptable exposure levels for allergic consumers. The establishment of population-based reference doses, such as the ED05 or VITAL thresholds, enables a scientifically sound basis for allergen risk assessment, linking analytical results with consumer protection goals [8]. This principle has been internationally endorsed by the FAO/WHO Expert Consultation on Risk Assessment of Food Allergens, which recommended health-based reference doses (RfDs) derived from probabilistic dose-distribution modelling as a foundation for evidence-based allergen management [9]. These developments align closely with the preventive intent of ISO 22002-100:2025, which emphasizes documented verification and validated analytical methods within PRPs.

2.2. *Immunochemical and Molecular Techniques for Allergen Detection*

A range of non-chromatographic analytical techniques complement chromatographic approaches in allergen detection, providing rapid and sensitive tools for both qualitative screening and quantitative determination. Enzyme-linked immunosorbent assays (ELISAs) remain the most established methods for routine monitoring, combining high specificity and sensitivity through antigen–antibody interactions and adaptable assay formats such as

https://doi.org/10.3390/separations13010039

sandwich and competitive ELISA [10]. Lateral flow devices (LFDs) extend this principle into portable, on-site applications, offering semiquantitative results within minutes and facilitating allergen verification along production lines [11]. More advanced optical thin-film biochips employ multiplexed detection techniques capable of simultaneously identifying several allergens through refractive index-based signal transduction, enabling miniaturized and high-throughput analyses [12]. DNA-based polymerase chain reaction (PCR) methods serve as confirmatory tools when protein integrity is compromised by processing, as they target stable genetic markers specific to allergenic species [13]. Finally, IgE antibody-based assays, including singleplex and multiplex immunoassays, link analytical outcomes with clinical relevance by detecting allergenic epitopes recognized by human sera [14]. Together, these immunochemical and molecular tools provide the foundation for effective allergen surveillance, supporting compliance verification and consumer protection across the food supply chain.

2.3. Chromatographic and Spectrometric Techniques in Allergen Management

Chromatographic and spectrometric techniques are fundamental to allergen management, supporting both traceability and quantitative verification across the food supply chain. Their use provides a robust scientific basis for evidence-based risk assessment and compliance validation under current food safety standards. Liquid chromatography–tandem mass spectrometry (LC-MS/MS) has become the reference analytical method for allergen determination, allowing simultaneous identification and quantification of allergenic peptides in complex matrices. Targeted proteomics using multiple reaction monitoring (MRM) and high-resolution mass spectrometry (HRMS) offers confirmatory protein identification, reinforcing validated traceability frameworks within food safety systems [15].

Advances in proteomics have enhanced allergen characterization using untargeted LC-MS analysis, data-independent acquisition (DIA), and label-free quantification, enabling precise peptide-level quantitation and detection of processing-induced modifications that affect allergenicity [16]. LC-MS-based proteomics now surpasses immunoassays in selectivity and robustness, combining reversed-phase and ion-exchange chromatography with electrospray ionization (ESI) or matrix-assisted laser desorption/ionization (MALDI) for sensitive peptide fingerprinting. The use of stable isotope-labelled standards further ensures quantitative reproducibility and method validation [17]. Despite advances in biosensor technologies, LC-MS remains the reference technique for confirmatory allergen analysis, providing high specificity, multiplexing capability, and robust structural characterization. Integration with miniaturized and automated microsystems represents a step toward real-time allergen surveillance [18].

Recent advances in LC-MS methods highlight the importance of standardization and matrix-aware quantification. Targeted selected and multiple reaction monitoring (SRM/MRM) methods can quantify allergenic proteins in complex food with sensitivity comparable to or exceeding ELISA, while offering multiplexing and structure-based confirmation [19]. The adoption of proteotypic peptides as reference analytes, supported by curated databases such as the Allergen Peptide Browser, improves method reproducibility and inter-laboratory consistency [20]. Extraction efficiency and peptide preparation, through defatting, enzymatic digestion, and desalting, remain critical to reliable quantification, particularly in lipid-rich or thermally processed products. Matrix components such as polyphenols and metal ions may alter protein solubility and epitope exposure, but advances in ultrasound- and microwave-assisted extraction mitigate these effects and enhance recovery [21].

https://doi.org/10.3390/separations13010039

Integration of systems biology and bioinformatics into MS-based allergenomics has broadened understanding of allergen structure-function relationships, enabling identification of clinically relevant peptide biomarkers and post-translational modifications [22]. Overall, LC-MS/MS has become the reference method for confirmatory allergen analysis and a key tool for validated analytical verification under ISO 22002-100:2025 requirements. Table 2 provides a comparative overview of analytical methods for allergen detection, summarizing their principles, main advantages, and limitations. It illustrates the complementary role of chromatographic methods alongside immunochemical and molecular assays in achieving reliable allergen quantification.

Table 2. Comparison of analytical techniques used for food allergen detection and quantification.

Method	Principle	Performance Characteristics	Limitations	Typical Use
ELISA	Specific antigen–antibody binding	High sensitivity, good reproducibility, easy to perform	Possible cross-reactivity, limited multiplexing	Official controls, HACCP verification, compliance testing
LFIA	Immunochromatographic reaction on membrane strips	Rapid results, portable format, suitable for on-site use	Mainly qualitative or semi-quantitative	On-site regulatory screening, hygiene and labelling verification
PCR	Amplification of species-specific DNA sequences	High specificity for source identification	Does not directly measure allergenic proteins	Traceability and compliance with EU allergen labelling requirements
LC-MS/MS	Detection of allergen-derived peptides after digestion	High specificity and sensitivity, multi-target capability	High instrumentation cost, complex sample preparation	Confirmatory analysis in official control and reference laboratories
MALDI-MS	Direct ionization and mass analysis of peptides	Fast analysis, low solvent consumption	Limited quantitative performance	Pre-screening in regulatory monitoring programmes
Biochip-based biosensors (optical or electrochemical)	Biological recognition combined with signal transduction	Miniaturization potential, integration into portable systems	Requires frequent calibration and validation	Supportive tools for regulatory surveillance and risk assessment

2.4. Validation, Standardization, and Harmonization of Chromatographic Methods for Allergen Analysis

Reliable allergen quantification depends on validated chromatographic methods that consistently perform across different food matrices and processing conditions. Validation confirms that the method is fit for purpose by assessing parameters such as specificity, recovery, precision (repeatability and reproducibility), accuracy, linearity, and limits of detection and quantification. Matrix-related factors (including extraction efficiency, peptide stability, and signal interference) must also be evaluated to ensure reliable measurement and traceable results. Standardization initiatives focus on defining common performance criteria, consistent reporting units, and calibration. Collaborative inter-laboratory studies coordinated by organizations such as AOAC, CEN/TC 275/WG12, and iFAAM aim to harmonize chromatographic and immunochemical methods, improving data comparability and ensuring that analytical performance aligns with clinically relevant allergen thresholds [23].

A major challenge in allergen analysis is the limited availability of well-characterized incurred certified reference materials (CRMs) for key allergens and food matrices. Existing CRMs, such as those developed for milk, egg, almond, hazelnut, and walnut, provide reliable materials for method validation, matrix-matched calibration, and recovery correction. They enable laboratories to assess inter-laboratory comparability and verify quantitative performance across analytical methods. However, the current CRMs cover only a small number of allergens and mainly simple or lightly processed matrices, which limits their applicability to more complex or heat-treated food [24]. Expanding CRM availability and implementing harmonized validation protocols are essential to ensure that chromatographic allergen measurements are scientifically reliable and suitable for evidence-based verification under ISO 22002-100:2025.

2.5. Targeted Chromatographic Applications for Individual Allergens

Targeted chromatographic applications have become essential tools for the accurate detection and quantification of specific food allergens. By focusing on characteristic peptides or proteins within defined food matrices, they achieve high selectivity and quantitative accuracy, even in thermally processed or complex food. Their application across individual allergen groups, including milk, egg, cereals, nuts, sesame, soy, fish, and mustard, demonstrates their importance for confirmatory allergen verification and compliance with food safety requirements.

For **milk allergens** (β-lactoglobulin, caseins), targeted LC-MS/MS enables reliable determination of heat-stable proteins and modified peptides in a wide range of processed matrices [25–30]. For **cereal allergens and gluten**, chromatographic and mass-spectrometric determination supports species-level discrimination and quantification of gliadins and glutenins below regulatory thresholds, including in thermally processed foods [31–33]. Analysis of **sesame allergens** benefits from targeting stable 2S, 7S, and 11S globulins, enabling accurate determination despite strong resistance to heat and digestion [34–36]. For **soy allergens**, peptide-based MS methods allow sensitive quantification of Gly m proteins in matrices where immunochemical assays often face limitations [37–40]. Chromatographic determination of **egg allergens** supports simultaneous identification of Gal d proteins and reliable assessment of processing effects on allergenicity [41–44]. Detection of **fish allergens**, especially parvalbumin, is strengthened by MRM-based LC-MS/MS, which minimizes interference arising from protein homology across fish species [45,46]. For **mustard allergens**, LC-MS/MS provides confirmatory detection of Sin a 1 in processed or acidic foods where ELISA and PCR methods may be affected by matrix interference [47,48]. Finally, targeted chromatographic applications for **peanut and tree nut allergens** allow sensitive determination of heat-stable proteins such as Ara h 3 and Jug r isoforms, supporting product validation and labelling accuracy [49,50].

A consolidated overview of all targeted chromatographic applications, including analytical techniques, matrices, and validation characteristics, is presented in Table 3. These studies highlight the selectivity and quantitative capability of peptide-based chromatographic determination and support confirmatory allergen verification in alignment with ISO 22002-100:2025 requirements. As individual allergen methods continue to advance in sensitivity and selectivity, analytical practice has increasingly shifted toward approaches capable of detecting multiple allergens simultaneously.

https://doi.org/10.3390/separations13010039

Table 3. Summary of targeted chromatographic methods for individual food allergens, including characteristic analytes, analytical techniques, food matrices, and key validation elements.

Ref.	Allergen(s)/Target Proteins	Matrix	Analytical Technique(s)	Sample Preparation Conditions	Key Findings/ Validation Notes
[25]	Milk proteins: Caseins, β-lactoglobulin, α-lactalbumin	Processed foods	LC-MS/MS	General extraction and peptide-level detection	Demonstrates need for highly sensitive and validated methods due to persistence of milk proteins after processing.
[26]	Milk peptides	Heat-treated and baked products	LC-MS/MS with immobilized trypsin	Rapid immobilized trypsin digestion enabling high peptide yield	High-throughput quantification, suitable for thermally processed matrices with efficient peptide release.
[27]	Hidden milk allergens	Multi-ingredient foods (e.g., meat products)	LC-ESI-HRMS	Standard protein extraction and HRMS profiling	Higher sensitivity and specificity than immunoassays, robust identification of hidden allergens in complex matrices.
[28]	β-lactoglobulin, αS1-casein, κ-casein, α-lactalbumin	Various food matrices	Multiplex LC-MS	Single-run multi-peptide quantification	Enables simultaneous determination of major milk allergens, high accuracy with reduced analysis time.
[29]	Milk proteins	Fruit-based or composite snacks	LC-MS/MS	Optimized digestion and solid-phase extraction and matrix-matched calibration	Highlights matrix interferences (polyphenols, pectins) reducing protein recovery, importance of tailored extraction strategies.
[30]	Milk proteins	Processed foods	LC-MS/MS	Optimized digestion and SPE and matrix-matched calibration	Detects milk < 5 mg/kg with high repeatability and recovery, suitable for confirmatory analysis and PRPs verification.
[31]	Gluten proteins (gliadins, glutenins)	Various cereal-based foods	General LC-MS/MS approach	Standard extraction and peptide profiling	Underscores need for sensitive/validated methods for gluten-free compliance due to very low reaction thresholds.
[32]	Gluten peptides, species differentiation (wheat, barley, rye)	Complex and processed cereal foods	LC-MS/MS	Robust extraction and peptide marker profiling	Capable of distinguishing cereal species, accurate quantification below the 20 mg/kg Codex gluten-free threshold.
[33]	Gluten peptides	Thermally processed foods	LC-MS/MS	Emphasis on peptide marker selection and matrix-matched calibration	Ensures reliable quantification in heat-treated matrices, highlights importance of optimized preparation and calibration to match matrix effects.

https://doi.org/10.3390/separations13010039

Table 3. *Cont.*

Ref.	Allergen(s)/Target Proteins	Matrix	Analytical Technique(s)	Sample Preparation Conditions	Key Findings/ Validation Notes
[34]	Gluten (qualitative/ semiquantitative)	On-site hygiene verification	TLC and colorimetric readout	Simple extraction and TLC migration	Portable screening tool with low detection limits (approximately 0.12 mg), suitable for rapid cross-contamination monitoring in industrial settings.
[35]	Sesame allergens (2S albumins, 7S vicilin-like, 11S globulins)	Various sesame-containing foods	General chromato-graphic/MS context	Standard extraction and protein/peptide detection	Highlights high heat and digestion resistance of major sesame allergens, underscores analytical challenges in processed foods.
[36]	Sesame storage proteins	Thermally processed foods	LC-MS/MS	Peptide biomarker selection for heat-stable proteins	Demonstrates difficulty of detecting sesame in processed matrices due to allergen stability, supports need for MS-based confirmatory methods.
[37]	Sesame peptide biomarkers (e.g., Ses i 1)	Diverse food matrices (including baked/fried/ boiled)	LC-MS/MS (targeted)	Extraction and optimized signature peptide selection	High specificity/sensitivity even after extensive processing, improved inter-laboratory consistency using robust peptide markers.
[38]	Soy allergens (general)	—	—	—	Background reference describing the regulatory importance of soy allergens and the need for accurate detection due to widespread use of soy-derived ingredients.
[39]	Soy proteins: Gly m 4, Gly m 5 (β-conglycinin), Gly m 6 (glycinin)	Various soy-containing foods	LC-MS/MS (targeted)	Protein extraction and peptide-level detection	Demonstrates high processing stability of key soy allergens, supports the need for MS-based confirmation where immunoassays show reduced performance.
[40]	Total soy protein (via stable isotope-labelled peptides)	Bakery and meat model matrices	Targeted LC-MS/MS	Enzymatic digestion and isotope-labelled internal standards	Enables precise quantification at very low levels (approximately 10 ppm), robust across isolates, concentrates, roasted and hydrolysed soy flours.
[41]	Soy residues in plant-based beverages and dairy alternatives	Complex matrices with denatured proteins	LC-MS/MS	Optimized enzymatic digestion and external calibration	Reliable detection in matrices where ELISA underestimates due to denaturation, supports PRP verification and regulatory compliance.

Table 3. *Cont.*

Ref.	Allergen(s)/Target Proteins	Matrix	Analytical Technique(s)	Sample Preparation Conditions	Key Findings/ Validation Notes
[42]	Egg allergens (Gal d 1–4: ovomucoid, ovalbumin, ovotransferrin, lysozyme)	Various food products	—	—	Background reference describing major egg allergens, their heat/enzymatic stability, and analytical challenges in processed/composite foods.
[43]	Gal d 1–4 (major egg white proteins)	Egg-containing foods, purified fractions	Ion-exchange chromatography	Protein extraction and IEC purification	Enables purification of four major allergens in a single step with high recovery and preserved immunological activity, supports allergenicity evaluation and method validation.
[44]	Gal d 1–4 (major egg white proteins)	Raw and baked egg matrices	LC-MS/MS	Extraction and optimized digestion and calibration	Demonstrates influence of thermal processing on allergen detectability, quantification requires matrix-specific calibration for accuracy.
[45]	Gal d 1–4 (major egg white proteins)	Fresh and heat-treated egg-containing foods	LC-MS/MS	Targeted peptide quantification	Provides trace-level detection (<10 mg/kg) with high reproducibility, suitable for analytical verification under ISO 22002-100:2025.
[46]	Fish allergens (parvalbumin)	Various fish-containing foods	—	—	Background reference describing parvalbumin as the main fish allergen, its high stability, and the analytical difficulty posed by species homology and processing.
[47]	Fish allergens (β-parvalbumin)	Fish-based products (including processed foods)	LC-MS/MS (MRM)	Tryptic digestion and mild denaturants to improve peptide recovery	Achieves sensitive quantification (<0.1 μg/g) with high precision and linearity, robust method suitable for routine allergen monitoring and verification of labelling accuracy.
[48]	Mustard allergens (Sin a 1)	Complex processed foods (sauces, bakery products)	LC-MS/MS (SRM)	Extraction and targeted peptide quantification	Highly specific and sensitive for Sin a 1 in processed matrices, detects sub-ppm levels (approximately 0.25 ppm), suitable for allergen traceability and ISO 22002-100:2025 verification.

https://doi.org/10.3390/separations13010039

Table 3. *Cont.*

Ref.	Allergen(s)/Target Proteins	Matrix	Analytical Technique(s)	Sample Preparation Conditions	Key Findings/ Validation Notes
[49]	Mustard allergens	Thermally treated or acidic foods	ELISA and PCR-based assays	Standard immunochemical/ molecular techniques	Describes limitations of ELISA/PCR due to matrix interference and thermal effects, supporting the need for confirmatory LC-MS/MS methods in mustard detection.
[50]	Peanut allergens (Ara h 3)	Processed nut-containing foods	DAS-ELISA and LFIA	Standard immunochemical extraction	Achieves highly sensitive detection (39 ng/mL), maintains precision and specificity after heat–moisture treatment, useful for monitoring processing-induced changes in antigenicity.

2.6. Simultaneous Chromatographic Determination of Multiple Food Allergens

Multiplex chromatographic approaches have become increasingly important for comprehensive allergen determination, as they allow several allergenic proteins to be measured within a single analytical sequence. By integrating common extraction and digestion steps with chromatographic separation and peptide-based quantification, these methods reduce analytical effort and improve consistency relative to single-target assays. When supported by appropriate validation, multi-allergen procedures offer robust trace-level detection across a wide range of food matrices and provide the analytical assurance required for verification of allergen control measures under ISO 22002-100:2025.

Applications of simultaneous LC-MS/MS determination demonstrated accurate quantification of combinations such as milk and egg allergens and later extended to nut proteins through peptide-based approaches targeting characteristic globulins [51,52]. Subsequent studies expanded multiplex analysis to priority allergens including milk, egg, and peanut, where optimized extraction, enzymatic digestion, and internal standardisation improved reproducibility and agreement with immunochemical methods [53–55]. Additional studies incorporating micro-HPLC, LC-MS, ELISA cross-comparison, and complementary optical techniques enabled concurrent detection of allergens such as casein, soy, and gluten in baked or composite foods [56–58]. More recent LC-MS/MS and LC-HRMS developments have achieved simultaneous determination of four or more allergenic proteins, including milk, egg, soy, peanut, crustaceans, and tree nuts, across processed and multi-ingredient matrices, with strong linearity, consistent recoveries, and inter-laboratory reproducibility [59–62]. Foundational studies further established the feasibility of detecting up to seven allergens within a single chromatographic run, while newer HRMS applications extend coverage to gluten and seafood allergens with high analytical accuracy [63–66]. A consolidated overview of simultaneous chromatographic applications (including analyte coverage, analytical techniques, matrices, and validation characteristics) is presented in Table 4.

https://doi.org/10.3390/separations13010039

Table 4. Summary of simultaneous (multiplex) chromatographic methods for the determination of multiple food allergens, including analyte coverage, analytical approaches, representative matrices, and validation characteristics.

Ref.	Allergen(s)/ Target Proteins	Matrix	Analytical Technique(s)	Sample Preparation Conditions	Key Findings/ Validation Notes
[51]	Milk, egg allergens	Composite and processed foods	LC-MS/MS with isotope-labelled internal standards	Extraction and unified digestion protocol	Accurate quantification of both allergen classes, detection < 1 mg/kg, excellent repeatability, demonstrates robust dual-allergen determination.
[52]	Nut allergens (walnut, almond proteins)	Diverse food matrices	LC-MS/MS (targeted)	Extraction and selection of 2S, 7S, 11S globulin peptides	High linearity (>0.999) and recoveries > 90%, demonstrates reliable simultaneous nut allergen identification and quantification.
[53]	Milk, egg, peanut allergens	Thermally processed foods	LC-MS/MS	Optimized extraction, enzymatic digestion, targeted peptide monitoring	LOQ ~5 mg/kg, enables simultaneous quantification of three priority allergens in complex heat-treated matrices.
[54]	Multiple allergens (milk, egg, peanut)	Processed foods	LC-MS/MS with matrix-matched calibration	Extraction, digestion, calibrants using allergen ingredients, isotope-labelled internal standards	Matrix-matched calibration significantly improves accuracy, strong reproducibility across matrices.
[55]	Multiple allergens (milk, egg, peanut)	Processed foods	LC-MS/MS and ELISA comparison	Standard extraction, peptide-based quantification	Good correlation with ELISA, suitable for verifying thermal processing effects and supporting allergen-labelling compliance.
[56]	Milk, egg, soy allergens	Baked food matrices	Micro-HPLC and dual-cell linear ion trap MS	Sonication-assisted extraction, tryptic digestion	Highly sensitive detection in baked matrices, LOD 0.1–2 µg/g, enhanced signal intensity due to optimized sonication, suitable for multi-allergen determination in thermally processed foods.
[57]	Casein, soy, gluten	Incurred cookies (baked products)	LC-MS and ELISA and multiplex flow cytometry	Extraction, digestion, multimethod comparison	Consistent detection across techniques, recoveries > 80% after baking, supports robustness of combined LC-MS with immunochemical tools for processed matrices.
[58]	Various food allergens including seafood proteins	Mixed food matrices	LC-MS/MS (reviewed approaches)	—	Review highlighting the expansion of multi-allergen LC-MS/MS strategies to additional allergenic protein groups (e.g., seafood), emphasizes potential for comprehensive allergen profiling.

https://doi.org/10.3390/separations13010039

Table 4. *Cont.*

Ref.	Allergen(s)/ Target Proteins	Matrix	Analytical Technique(s)	Sample Preparation Conditions	Key Findings/ Validation Notes
[59]	Milk, egg, soy, peanut allergens	Processed foods (cookies, sauces, chocolate)	UHPLC-MS/MS	Extraction, enzymatic digestion, targeted peptide quantification	LOQ < 5 mg/kg, high linearity in incurred samples, reliable multi-allergen detection across different processed foods.
[60]	Milk, egg, soy, crustacean peptides	Fish and meat products	LC-HRMS	Extraction, digestion, high-resolution identification	Recoveries 80–95%, strong agreement with ELISA screening, robust simultaneous detection across mixed matrices.
[61]	Milk, egg, peanut, hazelnut allergens	Mixed food matrices	Quantitative LC-MS	Standard extraction, enzymatic digestion, use of internal standards	Demonstrated strong inter-laboratory reproducibility, suitable for harmonized multi-allergen verification protocols.
[62]	Milk, egg, peanut, hazelnut allergens	Various processed foods	LC-MS/MS	Generic extraction, digestion	High consistency across runs, confirms applicability of LC-MS/MS for harmonized multi-allergen determination.
[63]	Milk, egg, soy, hazelnut, peanut, walnut, almond	Incurred bread samples	LC-MS/MS	Unified extraction, digestion, single-run targeted MS	First demonstration of simultaneous detection of 7 allergens in one run, quantification range 10–1000 µg/g, feasibility of multiplex allergen LC-MS/MS established.
[64]	Multiple allergens	Processed foods	LC-MS/MS (enhanced conditions)	Optimized extraction, digestion, refined chromatographic setup	Improved peptide recovery vs. Ref. [60] and throughput, better performance in processed matrices, advances in multiplex method robustness.
[65]	Multiple allergens including gluten and crustacean proteins	Diverse processed food matrices	LC-HRMS	High-resolution identification, targeted quantification	Accurate simultaneous quantification, recoveries 70–110%, extended scope beyond classical priority allergens.
[66]	Multiple allergens including gluten and crustaceans	Mixed food products	LC-HRMS	Standardized extraction, HRMS acquisition	High accuracy across complex matrices, confirms HRMS suitability for expanded multi-allergen suites, aligns with needs for broad-spectrum verification.

These analytical methods provide the sensitivity, selectivity, and traceability required for the verification activities specified in ISO 22002-100:2025, ensuring that allergen control measures can be confirmed with validated methods. Their capacity to deliver harmonized quantitative data for several allergenic proteins within a single chromatographic analysis strengthens risk-based allergen assessment, supporting objective evaluation of cross-contact and enabling scientifically justified decisions regarding precautionary and mandatory allergen labelling.

https://doi.org/10.3390/separations13010039

3. Chromatographic Approaches for Food Fraud and Authenticity Control

3.1. Regulatory and Management System Standards Requirements for Food Fraud Control

Food fraud and authenticity assurance are shaped by a defined set of regulatory and management system requirements in the European Union (EU) and internationally. Regulation (EC) No 178/2002 establishes consumer protection, non-misleading practices, and traceability as core legislative obligations, forming the basis for authenticity across the supply chain [67]. Regulation (EU) 2017/625 further requires competent authorities to implement risk-based controls addressing integrity, fair trade practices, and potential consumer deception related to origin, composition, or production method [68]. Together, these regulations position authenticity as a legal requirement and support the use of analytical testing, including chromatographic approaches, for compliance verification.

ISO 22000:2018 incorporates risk-based thinking at both organizational and operational levels, requiring food businesses to identify and manage risks that may affect the performance of the food safety management system [3]. Although the standard primarily targets food safety hazards, its structure permits the integration of food fraud–related vulnerabilities through requirements on traceability, control of externally provided materials, and verification activities. The newly published ISO 22002-100:2025 further strengthens this approach by introducing a dedicated Food Fraud Mitigation Plan, systematic vulnerability assessment, reinforced supplier approval and traceability expectations, and explicit requirements for analytical verification of authenticity [1]. As the first cross-sectoral PRP to address food fraud in a structured manner, it provides a harmonized basis for linking supply-chain controls with laboratory-based authenticity verification.

Core definitions, typologies, and mechanisms of food fraud are well established in the scientific literature, providing the basis for identifying fraud types and their implications for food chain integrity [69]. Foundational analyses further document recurring commodity-specific vulnerabilities, characteristic adulteration patterns, and the enduring public health and economic impacts that shape current prevention requirements [70]. Recent evidence reinforces the need for structured, vulnerability-based approaches to food fraud mitigation. Consumer-oriented analyses show that effective countermeasures depend on proactive, data-driven identification of fraud risks and on transparency and alignment between regulatory and industry controls [71]. Holistic vulnerability-assessment approaches incorporate multiple indicators, including product characteristics, historical fraud patterns, and supply-chain complexity, supporting the systematic assessment processes now embedded in ISO 22002-100:2025 [72]. EU agri-food fraud surveillance data demonstrate the practical implementation of Regulation (EU) 2017/625, with recurring fraud patterns detected through market controls, border inspections, and operators' own checks [73]. Long-term global fraud records likewise show persistent vulnerabilities in high-value commodities such as milk, olive oil, honey, meat products, and spices, underscoring the need for structured mitigation plans within management system standards [74].

These regulatory and standards-based requirements create the prerequisite conditions for effective food fraud mitigation, which increasingly relies on structured prevention models and analytical verification tools.

3.2. Mitigation Strategies for Food Fraud Prevention

Food fraud mitigation has shifted from reactive incident response to structured, prevention-oriented strategies based on risk and vulnerability assessment. Core principles emphasize an iterative cycle of vulnerability identification, risk prioritization, targeted control implementation, and continuous verification [75]. Effective prevention requires coordinated governance, integration of relevant intelligence, and focused controls across procurement, specifications, and supply-chain management. Reviews of major mitigation

https://doi.org/10.3390/separations13010039

guides consistently highlight four prevention elements, reducing motivation, limiting opportunities, improving detection, and reinforcing deterrence, operationalized through supplier controls, specification requirements, risk-based analytical testing, and market-surveillance activities [76].

Recent analyses of food fraud mitigation guidance documents underscore the importance of consistent terminology and clearly structured guidance, as definitional clarity strengthens both vulnerability assessment and the selection of proportionate preventive measures [77]. Complementary research on anticounterfeiting solutions highlights the expanding role of technological tools, such as overt and covert security features, tamper-evident packaging, digital identifiers, and enhanced traceability systems, in supporting product authentication and reinforcing supply-chain integrity, especially for high-value commodities [78]. Together, these developments complement management system controls and contribute to more robust, prevention-focused fraud-mitigation schemes.

Additional evidence highlights contextual and operational factors that shape the effectiveness of food fraud mitigation. Studies note that structural constraints, such as limited analytical capacity, insufficient data, and resource shortages, can hinder implementation, particularly in low-income settings, underscoring the need for scalable controls and improved information sharing across the supply chain [79]. At the same time, comparative evaluations of Vulnerability Assessment and Critical Control Points (VACCP) approaches show that the lack of a harmonized methodology produces variability in how vulnerabilities are assessed and controls prioritized [80]. Greater alignment on assessment steps, data inputs, and prioritization criteria is increasingly recognized as essential for ensuring risk-based, consistent, and globally comparable mitigation practices.

GFSI has played a critical role in formalizing industry-wide expectations for food fraud prevention, progressively moving from general authenticity requirements to clearly defined obligations for vulnerability assessment, documented mitigation plans, and evidence-based verification. Over the past decade, GFSI-recognized certification schemes have increasingly required structured, risk-driven controls that integrate supplier assurance, specification management, incident monitoring, and targeted analytical verification [81]. These developments have influenced the structure of ISO 22000-based systems and have contributed to the explicit food fraud requirements included in ISO 22002-100:2025.

In parallel, food defence has become a key complementary component of current management system standards. Through structured Threat Assessment and Critical Control Points (TACCP), organizations are expected to assess and manage intentional, malicious contamination threats, whether arising from disgruntled employees, extortion attempts, ideological actors, cyber-attacks, or broader terrorism-related risks. Effective food defence plans rely on clear threat characterization, identification of vulnerable points, proportionate security measures, and regular verification to ensure ongoing effectiveness [82]. Although not directly connected to analytical chromatography, food defence reflects the broader move toward integrated integrity systems in which food safety, fraud prevention, and threat management function as interconnected elements of supply-chain resilience.

Recent evaluations highlight that analytical testing serves as a targeted verification tool within food fraud mitigation rather than a primary detection mechanism. Most fraud events are not identified through routine laboratory testing due to sampling limitations, cost constraints, and the variable nature of adulteration practices. Analytical methods are therefore applied selectively and require fit-for-purpose sampling plans, validated methods, and appropriate detection capabilities [83]. In this context, chromatographic and mass-spectrometric methods operate as risk-based confirmatory tools that strengthen preventive controls and support the validation of authenticity claims.

https://doi.org/10.3390/separations13010039

3.3. Chromatographic Strategies for Food Fraud Detection and Authenticity Assessment

Chromatography is widely used in food fraud detection because it enables selective separation and reliable identification of compounds linked to authenticity. It supports targeted analysis of known markers or adulterants and non-targeted profiling, where LC-MS or GC-MS data are compared with authenticated reference samples. Targeted approaches confirm specific fraud scenarios, whereas non-targeted methods detect unexpected compositional deviations. The integration of high-resolution mass spectrometry and chemometric data analysis further improves sample classification and strengthens authenticity assessment [84].

Over the past five years, chromatographic techniques have been increasingly complemented by modern data-analysis tools that enhance the interpretation of complex analytical results. Strategies combining LC-MS and GC-MS with advanced computational models allow more efficient detection of compositional deviations and subtle indications of adulteration. Improvements in automated processing, faster separation procedures, and simplified evaluation steps have reduced analysis time without compromising analytical reliability, supporting broader application across diverse food categories [85].

Recent reviews highlight the rapid expansion of chromatographic and MS-based techniques used for authenticity assessment. Systematic evaluations report growing integration of LC-MS, GC-MS, isotope ratio methods, and other hyphenated approaches with chemometric modelling to improve the detection of adulteration, origin misrepresentation, and species substitution [86]. Beyond conventional approaches, multi-modal analytical methods that combine chromatographic separation with complementary spectral or molecular techniques yield more robust compositional information for authenticity verification [87].

Further studies show that combining chromatographic data with spectroscopic or DNA-based techniques enhances detection capability, particularly in complex matrices. Multimodal chemometric fusion models increase classification accuracy and reduce false detections, while advanced statistical classification of chromatographic data enables rapid screening and supports the early recognition of abnormal compositional patterns indicative of adulteration [88,89].

Despite these advances, a notable gap remains between academic developments and methods routinely implemented in industry. Comparative assessments reveal that although high-resolution untargeted LC-MS and GC-MS approaches can detect subtle deviations, industrial laboratories often rely on simpler targeted methods due to cost, throughput, and expertise constraints [90]. At the same time, there is increasing emphasis on rapid screening solutions, such as LC-MS-based triacylglycerol profiling, streamlined MS evaluations, and multivariate classification, to support faster decision-making and reduce reliance on labour-intensive confirmatory testing. Collectively, these trends point to the need for analytically robust yet operationally feasible chromatographic tools for routine authenticity verification [91].

Volatile-metabolome analysis represents an important complementary area, with GC-MS enabling sensitive characterization of volatile organic compounds (VOC) reflective of raw materials, processing conditions, and storage history. Distinct VOC compositions can act as authenticity markers or indicators of tampering, while solvent-free sampling approaches such as solid-phase micro-extraction facilitate detection of trace-level components relevant to adulteration. Untargeted VOC profiling has demonstrated strong discriminatory ability across multiple food categories, underscoring its value in authenticity assessment [92].

High-performance thin-layer chromatography coupled with mass spectrometry (HPTLC-MS) has also emerged as a practical supplementary technique. It enables rapid separation of key constituents and direct MS verification of selected zones, offering an

https://doi.org/10.3390/separations13010039

efficient screening option for detecting adulteration in products with characteristic chemical markers [93].

Chromatographic and MS-based methodologies now constitute a technically mature tool for authenticity assessment. Their practical relevance becomes most evident when these methods are applied to real food matrices, where fraud patterns and compositional variability determine analytical performance.

3.4. Applications in Food Analysis

3.4.1. Overview Across Major Food Categories

Chromatographic techniques demonstrate their full analytical value when adapted to specific food matrices, where compositional complexity, characteristic adulteration pathways, and matrix-dependent interferences shape method selection and detection capability. Across major food categories, targeted and non-targeted LC-MS and GC-MS approaches supported by chemometrics are widely applied to verify identity, detect substitution, and strengthen traceability within risk-based control systems.

Organic foods. LC-MS/GC-MS compositional profiles combined with multivariate modelling reliably distinguish organic from conventional products and support verification of cultivation and origin claims, while advanced chemometric tools enhance detection of irregularities in the organic supply chain [94,95].

Beverages. Chromatographic methods and isotope ratio analysis enable authentication of wine, spirits, and beer by verifying botanical origin, geographic provenance, processing practices, and detecting dilution or exogenous components. Integrated approaches combining chromatography, spectroscopy, and stable isotopes address complex fraud mechanisms. Non-alcoholic beverages remain vulnerable to counterfeiting and adulteration, requiring targeted analytical testing within robust control systems [96–99].

Plant-origin foods. LC-MS, GC-MS, and isotope ratio methods support discrimination of origin, cultivar, and botanical identity in coffee, spices, herbs, grains, and cereals. Chromatographic, spectroscopic, and DNA-based methods improve detection of common adulterants, fillers, and misrepresentation of quality or geographic origin [100–103].

Oils and fats. Extra virgin olive oil (EVOO) authenticity assessment relies on chromatographic compositional characteristics, targeted markers, and MS-based profiling to detect blending and verify origin. Similar chromatographic and spectroscopic approaches are applied to other specialty oils, reflecting diverse adulteration risks across the category [104,105].

Honey. Due to compositional variability and globalized sourcing, honey authentication relies on combined chromatographic, isotopic, and spectroscopic techniques, including LC-IRMS, HRMS profiling, NMR fingerprinting, and targeted/non-targeted LC-MS and GC-MS approaches, to detect sugar-syrup dilution, verify botanical and geographical origin, and evaluate potential losses of functional bioactive constituents [106–108].

Meat, poultry, and seafood. High fraud susceptibility, through species substitution, dilution, or mislabelling, necessitates DNA-based identification supported by chromatographic, spectroscopic, and elemental analysis to verify species integrity and production claims [109,110].

Milk and dairy products. LC-MS and related chromatographic methods are central for detecting water dilution, foreign proteins or fats, species mislabelling, and cheese adulteration. Combined chromatographic, spectroscopic, and MS-based approaches, often chemometric-assisted, strengthen authentication of milk, dairy ingredients, and PDO/PGI cheeses [111–115].

For **PDO/PGI products**, stable isotope profiles can support verification of geographical and, in some cases, production-origin claims. From a chromatographic perspective,

https://doi.org/10.3390/separations13010039

compound-specific isotope analysis (CSIA) using GC-C-IRMS and/or LC-IRMS resolves isotopic information at the compound level, which is useful in complex matrices when markers such as sugars, amino acids, fatty acids, or selected aroma constituents are targeted. Bulk IRMS may still be used for screening purposes, while chromatographic CSIA is better suited to situations where confirmatory, compound-resolved evidence is required [116–118].

Recent reviews on synthetic **food colourant** analysis highlight LC-MS/MS (typically triple-quadrupole methods operated in selected reaction monitoring mode) as a confirmatory approach for targeted determination of both permitted dyes (compliance control) and undeclared/unauthorized synthetic colourants, supported by matrix-adapted sample preparation (e.g., solid-phase extraction, QuEChERS, microextraction) to mitigate interferences in complex foods [119].

To provide a structured synthesis aligned with the verification requirements introduced in ISO 22002-100:2025, Table 5 summarizes key fraud mechanisms across major food categories together with the chromatographic and complementary analytical techniques most frequently applied for authenticity assessment.

Table 5. Overview of chromatographic and related analytical techniques applied to food authenticity verification across key food categories.

Refs.	Food Category	Typical Fraud Mechanisms	Chromatographic/Analytical Approaches
[94,95]	Organic foods	Mislabelling of cultivation practices, substitution with conventional products	LC-MS/GC-MS fingerprints, multivariate chemometrics, compositional pattern recognition
[96–98]	Beverages (alcoholic)	Dilution, addition of exogenous alcohol, misrepresentation of origin/vintage	LC-MS/GC-MS profiling, trace-level markers, multi-isotope IRMS, combined chromatography–spectroscopy techniques
[99]	Beverages (non-alcoholic)	Counterfeiting, juice adulteration, undeclared additives	Targeted chromatographic assays, quality system-based verification, complementary spectroscopic screening
[100]	Plant-origin foods (general)	Substitution, dilution, misrepresentation of botanical/geographical origin	LC-MS/GC-MS, isotope ratio analysis, chemometric classification
[101,102]	Coffee, spices, herbs	Addition of fillers, undeclared plant materials, colorants, flavour enhancers	Chromatographic profiling, MS-based detection, spectroscopic markers, DNA assays
[103]	Grains and cereals	Varietal substitution, protein inflation, false origin/production claims	LC-MS/GC-MS, spectroscopic profiling, DNA assays, IRMS
[104,105]	Oils and fats	Blending with cheaper oils, misrepresentation of botanical/geographic origin	Fatty acid profiling, sterol analysis, MS-based fingerprints, chemometrics
[106–108]	Honey	Sugar-syrup dilution, false botanical or geographical origin, post-processing manipulation, loss of functional bioactive constituents	LC-IRMS, HRMS profiling, NMR fingerprinting, targeted and non-targeted LC-MS/GC-MS analysis

https://doi.org/10.3390/separations13010039

Table 5. *Cont.*

Refs.	Food Category	Typical Fraud Mechanisms	Chromatographic/Analytical Approaches
[109]	Meat and poultry	Species substitution, dilution, misrepresentation of production attributes	Chromatographic profiling, spectroscopic screening, DNA-based species authentication
[110]	Seafood	Species substitution, false origin/method of production, short-weighting	Molecular identification, LC-MS/GC-MS, elemental and isotope ratio markers
[111,114]	Milk	Water dilution, addition of foreign proteins/fats, false species origin	LC-MS/HPLC profiling of proteins and peptides, detection of nitrogen-rich adulterants
[112,115]	Cheese	Substitution with cheaper milk types, dilution, false PDO/PGI claims	LC-MS profiling, MIR/NIR/Raman spectroscopy, MS-assisted screening with chemometrics

3.4.2. Analytical Evidence from Individual Studies Across Food Matrices

Chromatographic, isotopic, and mass-spectrometric methods have been applied across major food matrices to quantify compositional markers, identify adulterants, and support origin and species verification. The following examples summarize representative applications based on targeted analytical parameters and experimentally validated chemical characteristics.

Beverages. LC-MS metabolomics has been used to distinguish grape varieties and production regions through quantitative metabolic profiles, while combined isotopic and chromatographic analyses detect dilution, C4-sugar enrichment, and synthetic additives in wine. In fruit juices, HPLC carbohydrate profiling and GC-MS volatile analysis enable detection of syrup addition and non-declared juice substitution, with multivariate analysis providing clear group separation [120–124].

Fruits and Vegetables. HS-SPME-GC-MS volatile profiling differentiates citrus cultivars and storage conditions, GC-C-IRMS separates natural from synthetic aroma constituents in apples, and chiral GC-MS identifies maturity and logistic pathways in pineapple. GC-MS/LC-MS analysis of grape volatiles and glycosides supports cultivar authentication, while chromatographic and colloidal profiles enable classification of processed tomato products and combined compositional parameters support discrimination of cherry and carrot geographical origin, with HPLC-DAD also supporting verification of natural colourant claims in relevant processed foods [125–132].

Cereal and Grain Products. Targeted lipidomics quantifies adulterated rice admixtures, untargeted LC-MS metabolomics separates organic from conventional rice, and proteomic peptide markers distinguish wheat, rye, and spelt in cereal products. LC-based phenolic profiling further differentiates organic and conventional wheat across successive harvests [133–136].

Vegetable and Seed Oils. NIR spectroscopy, TGA-GC-MS, and GC-MS volatile profiling detect undeclared seed-oil addition in camellia, olive, and sesame oils, while combined GC-MS analysis and hyperspectral data enable discrimination of adulterants in safflower oil. LC-MS glyceride analysis characterizes TAG/DAG distributions in fruit seed oils to verify botanical identity and identify compositional deviations indicative of adulteration [137–143].

Honey. HPLC-UV carbohydrate profiling quantifies C3/C4 syrup additions, HPLC organic-acid analysis differentiates monofloral honeys, and multi-wavelength HPLC-UV fingerprinting separates honeys of different geographical origin. These chromatographic approaches detect adulteration and origin mislabelling with higher resolution than routine physicochemical testing [107,108,144–146].

https://doi.org/10.3390/separations13010039

Meat and Fish. LC-MS/MS proteomics identifies species through proteotypic peptides in raw and processed meats, while targeted peptide markers discriminate bovine from non-bovine matrices. In fish, LC-HRMS profiling of muscle proteins provides species-specific peptide markers enabling discrimination of closely related species [147–149].

Milk and Dairy Products. SFC-Q-TOF-MS and RP-HPLC TAG profiling differentiate milk from multiple animal species and detect non-milk fat addition, while LC-MS/MS peptide profiling and MALDI-TOF-MS protein spectral characteristics authenticate cheeses and reveal undeclared milk sources. UHPLC-ELSD/UV chromatographic analysis enables detection of fat adulteration in butter, while RP-HPLC peptide profiling combined with MIR spectroscopy supports classification of cheeses according to geographical origin [150–155].

Table 6 summarizes representative studies across major food categories in alignment with the verification requirements introduced in ISO 22002-100:2025, with emphasis on documented analytical evidence, multi-marker validation, and matrix-specific vulnerability assessment. The table consolidates key fraud mechanisms, chromatographic and MS-based analytical approaches, diagnostic markers, and reported performance characteristics, providing a comparative overview of method performance under practical testing conditions.

Table 6. Representative chromatographic and MS-based applications for food fraud and authenticity assessment across major food matrices.

Ref.	Food Category	Food Matrix	Fraud/Authenticity Objective	Analytical Technique(s)	Target Analytes/ Diagnostic Markers	Reported Analytical Performance
[120]	Beverages	White wine	Varietal and geographical origin	LC-MS metabolomics	metabolic profiles	Clear discrimination of Greek varieties, robust PCA/PLS-DA separation
[121]	Beverages	Red wine	Dilution, C4 sugar chaptalization, synthetic additives	Stable isotope ratio analysis ($\delta^{13}C/\delta^{18}O$) and LC/GC compositional profiling	isotopic ratios, HMF, sweeteners, anthocyanins	Differentiation of authentic/ suspicious/adulterated samples
[122]	Beverages	Wine and fruit juices	Dilution, syrup addition, substitution, mislabelling	LC/GC profiling and IRMS and chemometrics	multi-marker chemical signatures	Review evidence supporting multi-approach strategies
[123]	Beverages	Apple juice concentrate	Syrup adulteration	HPLC RID and chemometrics	glucose/fructose ratio, maltose, sorbitol shifts	Detection $\geq$ 10% adulteration
[124]	Beverages	Citrus juices	Mandarin over-blending	HS-SPME-GC-MS VOC profiling	monoterpenes, esters, aldehydes	Clear PCA separation, detection $\geq$ 10%
[125]	Fruits	Oranges	Organic vs. conventional, cultivar/storage effects	HS-SPME-GC-MS	terpenes and esters	Distinct volatile profiles for organic fruit
[126]	Fruits	Apples	Synthetic vs. natural aroma	GC-C-IRMS	$\delta^{13}C$ of 16 marker volatiles	Accurate detection of synthetic additions
[127]	Fruits	Pineapple	Maturity stage and logistic history	Chiral HS-SPME-GC-MS	γ-/δ-lactone enantiomers	Differentiation of air- vs. sea-freighted fruit
[128]	Fruits	Grapes	Cultivar identification	GC-MS and LC-MS	monoterpenes, norisoprenoids, glycosides	Stable cultivar-associated chemical profiles

https://doi.org/10.3390/separations13010039

Table 6. *Cont.*

Ref.	Food Category	Food Matrix	Fraud/Authenticity Objective	Analytical Technique(s)	Target Analytes/ Diagnostic Markers	Reported Analytical Performance
[129]	Vegetables/ processed food	Tomato sauce	Origin/brand traceability	GC-FID/GC-IMS and asymmetric flow field-flow fractionation	volatile and colloidal profiles	High discrimination with chemometrics
[130]	Vegetables/ processed food	Blue confectionery	Verification of natural colourant claim	HPLC-DAD	C-phycocyanin (from *Arthrospira platensis*)	Validated method applied to commercial samples suitable for targeted verification of product claims
[131]	Vegetables	Sweet cherry	Botanical and geographical origin	HPLC sugars and GC-MS volatiles and ICP-OES minerals	multi-parameter profiles	>95% correct classification
[132]	Vegetables	Carrots	Geographical origin	Untargeted UHPLC-HRMS	region-specific metabolites	Strong origin discrimination, weaker for production mode
[133]	Cereals	Rice	Detection of adulterated admixtures	Targeted lipidomics (LC-MS based) and chemometric classification	lipid classes, molecular species	High accuracy classification of blends
[134]	Cereals	Rice	Organic vs. conventional	Untargeted LC-MS metabolomics	discriminant metabolites	Clear metabolic separation via PCA/PLS-DA
[135]	Cereals	Bread	Wheat/rye/spelt species ID	Targeted proteomics	species-unique peptides	Reliable detection of grain-type substitution
[136]	Cereals	Wheat	Organic vs. conventional	LC-based phenolic profiling	protocatechuic acid	Consistent differentiation across harvest years
[137]	Oils	Camellia oil	Seed oil adulteration	NIR spectroscopy and PLS-DA/PCA	lipid overtone absorption	>95% accuracy across mixtures
[138]	Oils	Olive oil	Soybean oil adulteration	TGA-GC-MS	thermal/volatile markers	Strong discrimination via temperature-resolved MS
[107]	Honey	Mixed botanical honeys	Syrup adulteration, floral origin	HPLC sugars, GC-MS/LC-MS volatiles, IRMS	$\delta^{13}C/\delta^{2}H$, sugar profiles	Strong adulteration detection, origin classification
[108]	Honey	Honey (general)	Functional degradation	LC-MS/LC quantification of phenolics, flavonoids, MGO	bioactive markers	Identifies heat/processing-induced loss of activity
[139]	Oils	Milk (oil adulteration)	Detection of added vegetable oils	Flash-GC and chemometrics	VOC profiles	High accuracy quantification despite matrix complexity
[140]	Essential oils	Mentha/ Ocimum	Adulteration with vegetable oils	FT-NIR and GC-MS/GC-FID and chemometrics	volatile chemical profiles	Detection at 3–30% adulteration
[141]	Seed oils	Safflower oil	Adulteration with multiple oils	GC-MS and hyperspectral imaging	sterols, fatty acids, and spectra	Fusion model improves discrimination

https://doi.org/10.3390/separations13010039

Table 6. *Cont.*

Ref.	Food Category	Food Matrix	Fraud/Authenticity Objective	Analytical Technique(s)	Target Analytes/ Diagnostic Markers	Reported Analytical Performance
[142]	Seed oils	Sesame oil	Undeclared seed oil blends	GC-MS and chemometric classification	volatile and lignan markers	Sensitive detection of low-level adulteration
[143]	Seed oils	Fruit seed oils	Botanical identity and adulteration	LC-MS glyceride profiling	TAG/DAG compositional profiles	Clear separation of oil types, non-authentic profiles detectable
[144]	Honey	Honey (adulter-ation trials)	C3/C4 syrup addition	HPLC-UV carbohydrates and PCA/PLS-DA	mono/di/ oligosaccharides	Clear grading across dilution levels
[145]	Honey	Robinia honey	Authentic vs. blended	HPLC organic acids	malic, citric, gluconic acids	Robust classification of authentic samples
[146]	Honey	Various origins	Geographical origin	Multiwavelength HPLC-UV	Spectral retention features	High accuracy regional discrimination
[147]	Meat	Multi-species meats	Species identification	LC-MS/MS proteomics (untargeted)	proteotypic peptides	Simultaneous detection in mixed/processed products
[148]	Meat	Beef	Species verification	Targeted LC-MS/MS peptides and chemometrics	bovine-specific peptides	Strong separation from non-beef matrices
[149]	Fish	Pollock vs. hake	Species identification	LC–HRMS protein-based analysis	species-specific peptide features	Accurate differentiation of closely related species
[150]	Dairy	Milk (multi-species)	Species authentication	SFC-Q-TOF-MS TAG profiling	TAG species	Clear species clustering (PCA/OPLS-DA)
[151]	Dairy	Milk/butter /cheese	Non-milk fat adulteration	RP-HPLC-RID TAG analysis	3 diagnostic TAG peaks	Detects 1–2% palm oil, strong chemometric separation
[152]	Dairy	Farmer's cheese	Authentic vs. industrial	LC-MS/MS peptides	endogenous peptide profiles	Distinguishes authentic/non-authentic samples
[153]	Dairy	PDO feta	Milk source adulteration	MALDI-TOF-MS	species-specific spectra	Rapid detection of non-sheep/goat milk
[154]	Dairy	Butter	Fat adulteration	UHPLC-ELSD and UHPLC-UV	pointwise chromatographic response profiles	Sensitive detection without markers
[155]	Dairy	Coalho cheese	Geographical origin	RP-HPLC peptides and MIR	peptide profiles and MIR spectral features	Accurate classification across 4 regions

The reported applications indicate that recently developed high-sensitivity chromatographic and MS-based techniques provide the analytical specificity required to address fraud vulnerabilities in high-risk food matrices. These approaches support industry implementation of evidence-based verification, as expected under ISO 22002-100:2025, while also strengthening regulatory-aligned controls for detecting economically motivated adulteration and managing authenticity risks.

https://doi.org/10.3390/separations13010039

4. Chromatographic Monitoring of Packaging Contaminants and Chemical Contamination in Food (Cleaning Agents, Disinfectants, Lubricants, and Pesticides Used for Pest Control)

Monitoring chemical contamination from packaging materials and auxiliary services in the food supply chain has become a critical requirement of prerequisite programmes (PRPs) under ISO 22002 100:2025. In line with ISO 22003-1:2022, the term auxiliary services encompasses activities such as cleaning and disinfection, pest control, maintenance, and other supporting operations that may represent indirect sources of chemical contamination in food processing environments. At the regulatory level in the European Union, general food hygiene obligations such as those set out in Regulation (EC) No 852/2004 require food business operators to establish, implement, and maintain effective cleaning, disinfection, pest control, and monitoring systems [156]. Meanwhile, specific regulations for materials intended to come into contact with food, such as Regulation (EC) No 1935/2004 and Regulation (EU) No 10/2011 on plastic materials, stipulate that migrations of constituents must not endanger human health nor lead to unacceptable changes in the food [157,158].

From an implementation perspective, guidance from industry and certification schemes further emphasizes the necessity of validated analytical verification of auxiliary chemical residues. For example, the GFSI guidance "Chemicals in Food Hygiene-Volume 2: Cleaning Agents, Sanitizers and Disinfectants in Food Businesses" provides a decision-tree approach for risk assessment of residues from cleaning and disinfection processes and highlights the need for precise detection methods, targeted sampling, systematic monitoring, and analytical verification [159].

Packaging materials, cleaning agents, disinfectants, lubricants, and pest control compounds are recognized sources of chemical contamination in food processing environments. Typical contaminants include migrant monomers and oligomers, NIASs, mineral oil hydrocarbons, quaternary ammonium compounds, peracetic acid by-products, glycols, and rodenticides. Chromatographic and spectrometric methods are therefore essential for verifying control effectiveness, ensuring regulatory compliance, and providing traceable analytical evidence in accordance with ISO 22002-100:2025.

4.1. Packaging-Related Chemical Contaminants

Food packaging materials represent a significant source of chemical contamination due to the migration of both intentionally and non-intentionally added substances into food. Migration involves monomers, oligomers, plasticizers, stabilizers, printing inks, adhesives, and varnish-derived components and is influenced by polymer structure, multilayer design, and storage conditions [160]. Contemporary food contact materials, including laminated plastics and biodegradable polymers, contain diverse classes of migrants such as impurities, degradation products, and reaction by-products, collectively classified as NIASs, which require selective chromatographic techniques for reliable detection [161]. The growing use of bio-based and biodegradable polymers adds further analytical complexity, as these materials may generate low-molecular-weight oligomers and transformation products that are not always covered by conventional toxicological datasets, increasing the need for non-target screening and high-resolution detection [162,163].

A wide range of chromatographic applications has been developed for the identification and quantification of packaging-derived contaminants, including volatile and semi-volatile migrants, oligomeric NIASs, photoinitiators, plasticizers, antioxidants, biocides, and mineral oil hydrocarbons across plastic, multilayer, paper-based, and recycled materials. These studies employ targeted and untargeted GC-MS/MS, LC-MS/MS, GC-GC/TOF-MS, HRMS, DI-SPME-GC-MS, LC-GC, and Orbitrap-based approaches, providing enhanced sensitivity for the characterization of complex migrant profiles, polymer-degradation

https://doi.org/10.3390/separations13010039

products, and MOSH/MOAH fractions. A consolidated overview of these analytical applications is provided in Table 7, which summarizes key compound classes, target analytes, matrices, and chromatographic methods reported across references [164–177].

These chromatographic advances directly support the verification requirements of ISO 22002-100:2025 by enabling sensitive, matrix-appropriate detection of packaging migrants, thereby strengthening evidence-based control of packaging-related chemical hazards.

4.2. Cleaning and Disinfectant Agent Residues

Cleaning agent residues represent a critical source of indirect chemical contamination in food processing environments, particularly in operations based on cleaning-in-place (CIP) systems, where alkaline, acidic, and oxidizing cleaning formulations are routinely applied to ensure hygienic conditions and the removal of organic and inorganic residues [178]. Conventional CIP performance indicators (e.g., visual inspection and conductivity-based rinse control) are insufficient to confirm the complete elimination of low-level chemical residues, supporting the need for validated, analytically based residue-monitoring strategies [178,179]. Risk-based cleaning procedures increasingly rely on systematic verification that combines optimized formulations with trace-level analytical control of post-cleaning surfaces to prevent hazardous carry-over into food matrices [179].

Dairy-processing environments represent a high-risk setting due to the intensive use of alkaline and chlorine-free cleaning agents in closed-loop CIP systems. Incomplete rinsing can result in the persistence of residues on product-contact surfaces, and chlorine-free cleaning practices have been shown to modify chemical residue profiles on equipment and pipelines, requiring dedicated analytical monitoring to ensure compliance with hygiene and safety requirements [180].

Various chromatographic applications have been developed for the determination of cleaning- and disinfectant-related residues, encompassing QACs, non-quaternary biocides, preservatives, disinfectants, cationic compounds, and surface-derived residues across dairy matrices, CIP-validation samples, rinse water, and stainless-steel surfaces. These methods employ UHPLC-MS/MS, LC-MS/MS, LC-HRMS, electromembrane extraction, magnetic SPE, modified stationary phases, and trace-level HPLC methods, enabling sensitive, selective, and regulatory-compliant residue determination. A consolidated overview of these analytical applications is provided in Table 7, which summarizes key compound classes, targeted analytes, matrices, and chromatographic approaches reported in references [181–190].

Overall, effective control of cleaning- and disinfectant-agent residues requires integration of risk-based cleaning design, validated sampling strategies, and sensitive chromatographic methods to ensure robust chemical-hazard control and compliance with PRP requirements.

4.3. Lubricant and Technical Fluid Contamination

Lubricants and technical fluids are widely used in food processing industries for the operation of equipment such as conveyors, bearings, compressors, hydraulic systems, and filling units and are recognized as potential indirect sources of chemical contamination when leaks, overapplication, or seal failures occur [191]. They consist of base oils blended with additives that provide thermal stability, load-bearing performance, and corrosion protection, characteristics that also make them relevant candidates for incidental food contact under industrial operating conditions [192].

Food-grade lubricants are formulated for applications where incidental food contact may occur and are subject to defined quality and safety criteria. Regulatory-oriented studies describe their classification, intended uses, and the importance of proper selection and handling to minimize contamination risks, particularly in high-throughput processing environments [193]. Increasing equipment complexity and extended production cycles further reinforce the need for analytical verification of lubricant-related residues as part of PRP implementation.

A comprehensive set of chromatographic applications has been developed for the determination of lubricant- and technical-fluid-derived residues across food, edible oils, packaging simulants, and food contact materials. These include normal-phase LC, GC-FID, HPLC-GC, SPE-GC, LC-MS/MS, GC-MS/MS, and LC-ESI-MS approaches for the characterization of MOSH/MOAH fractions, lipid-like migrants, degradation products, ester-oil transformation products, antioxidants, and additive profiles. Recent studies have also addressed the analytical challenges posed by biodegradable and natural oil-based lubricants, which generate polar hydrolysis- and oxidation-derived by-products requiring dedicated chromatographic detection beyond conventional residue methods. A consolidated overview of these applications is presented in Table 7, summarizing compound classes, target analytes, matrices, and analytical methodologies reported in references [194–202].

Overall, lubricant and technical fluid contamination remains an analytically demanding but important aspect of chemical risk management in food processing industries. The combined use of targeted chromatographic methods, advanced sample preparation techniques, and risk-based monitoring is essential for effective control of these indirect contaminants within PRP-based food safety systems.

4.4. Insecticide Residues from Indoor Pest Control

Indoor pest control treatments can leave insecticidal residues on equipment, structural surfaces, or stored foods within processing facilities. The compounds most commonly detected indoors belong to the synthetic pyrethroid and neonicotinoid classes, with diamides contributing to a smaller extent due to their targeted use in structural pest management programmes. This subsection focuses on insecticidal active substances used in structural pest management (pest control treatments) and the resulting residues in facility environments. In this context, pest control insecticides (e.g., insecticides for arthropod control) should be distinguished from disinfectants, which represent a separate category of sanitation-related substances.

Synthetic pyrethroids, characterized by low volatility, strong surface affinity, and high stability, persist on metal and polymeric materials and accumulate in dust and packaging, enabling secondary transfer to food products. Their determination relies mainly on GC-MS/MS and LC-MS/MS supported by microextraction techniques that reduce solvent use and enhance trace-level sensitivity [203]. Neonicotinoids such as imidacloprid, acetamiprid, and thiamethoxam may also be encountered in facility environments and show persistence together with stable metabolites (e.g., olefin, urea, hydroxy derivatives) detected on treated surfaces. SPE or QuEChERS extraction followed by LC-MS/MS or HRMS allows selective identification of parent compounds and metabolite residues at low μg/kg levels [204].

Diamide insecticides, including chlorantraniliprole and flubendiamide, represent a smaller but relevant group of indoor residues. Their polarity and structural diversity require clean-up prior to LC-MS/MS to minimize matrix interference [205]. For flubendiamide, both parent compound and the des-iodo metabolite can be reliably determined using LC-UV or LC-MS/MS [206].

Green microextraction approaches enhance the analytical capacity for pyrethroid determination. Hydrophobic deep-eutectic-solvent ferrofluids enable rapid magnetic preconcentration with GC-MS/MS detection at ng/mL levels [207]. Rhamnolipid-based bio-supramolecular solvents support low-solvent HPLC-UV analysis of pyrethroids in complex matrices [208]. Broader evaluations confirm the suitability of deep eutectic solvents across

https://doi.org/10.3390/separations13010039

multiple microextraction formats, enhancing selectivity and reducing solvent consumption in pesticide residue monitoring [209].

A broad spectrum of chromatographic applications has been developed to verify indoor pesticide residues across rodenticides, neonicotinoids, organophosphates, pyrethroids, and other insecticide classes relevant to food processing facilities. These include multi-residue determinations, chiral separations, micellar- and magnetic-based microextraction approaches, modified QuEChERS pretreatment techniques, dual-detector GC analyses, and high-resolution LC-MS methods applied to a wide range of commodities and contamination scenarios. Collectively, the studies reported in references [210–220] demonstrate the analytical diversity required to detect parent compounds, metabolites, stereoisomers, and degradation products at low concentrations, while also highlighting the influence of washing and processing steps on residue behaviour. A consolidated overview of these applications is presented in Table 7, which summarizes target analytes, matrices, and chromatographic methodologies relevant to indoor pest control verification.

These applications demonstrate the analytical capacity required to effectively monitor indoor-relevant pesticide residues. They further indicate that sensitive, low-solvent chromatographic approaches are essential for meeting the verification and control requirements established under ISO 22002-100:2025.

Table 7. Consolidated overview of chromatographic applications for the determination of packaging-related migrants, cleaning and disinfectant residues, lubricant- and technical-fluid-derived contaminants, and indoor pesticide residues relevant to PRPs under ISO 22002-100:2025.

Ref.	Class/Chemical Group	Target Analytes	Matrix	Analytical Technique(s)	Key Analytical Notes
Packaging-related chemical contaminants					
[164]	Multiple packaging-related migrants	Various emerging contaminants	Food contact materials	LC-MS/MS, GC-MS/MS, HRMS	Integration of targeted and untargeted approaches
[165]	Volatile and semi-volatile migrants	VOCs, SVOCs	Polypropylene containers → food simulants	GC×GC-TOF-MS	Reveals complex migration profiles
[166]	Bisphenols, phthalates, oligomeric NIAS	BPA/BPS, phthalates, oligomers	Infant food packaging	Targeted LC-MS/MS and suspect screening	Migration influenced by multilayer design and storage
[167]	Plasticizers	Phthalates, adipates, citrates	Various packaging plastics	GC	Migration dependent on fat content, temperature, contact time
[168]	Plasticizers	Phthalates	Packaging materials and foods	Microextraction and GC/LC	Improved preconcentration with reduced solvent use
[169]	Recycled plastic NIAS	Degradation and oxidation products	Recycled plastics	Non-target HRMS	Complex NIAS mixtures identified
[170]	Recycled material NIAS	Oxidation products, unknown migrants	Recycled plastics	HRMS, suspect screening	Supports identification of potentially hazardous NIAS
[171]	Isothiazolinone preservatives	Methylisothiazolinone, others	Food contact adhesives	LC-MS/MS	Selective quantification for routine compliance
[172]	Paper/cardboard migrants	Semi-volatile migrants	Fibre-based packaging	DI-SPME–GC-MS	Efficient screening under realistic contact conditions
[173]	Photoinitiators, bisphenols, plasticizers	Photoinitiators, bisphenols, phthalates, antioxidants, biocides	Paper and cardboard	GC-Orbitrap	Expanded non-targeted detection capabilities

https://doi.org/10.3390/separations13010039

Table 7. *Cont.*

Ref.	Class/Chemical Group	Target Analytes	Matrix	Analytical Technique(s)	Key Analytical Notes
Packaging-related chemical contaminants					
[174]	NIAS (paper-based)	Multiple NIAS classes	Recycled paper/cardboard	LC-HRMS/MS	High-resolution profiling of packaging migrants
[175]	Mineral oil hydrocarbons (MOH)	MOSH/MOAH	Cocoa products contaminated via jute bags/recycled packaging	LC-GC	Supply-chain contribution demonstrated
[176]	Mineral oil hydrocarbons	MOSH/MOAH	Packaging migrants	On-line LC-GC	Improved fractionation and reduced interferences
[177]	Mineral oil hydrocarbons	MOSH/MOAH	Routine surveillance	LC-GC-FI (with N_2 carrier)	Nitrogen as alternative carrier gas with acceptable performance
Cleaning and disinfectant agent residues					
[181]	Residual milk proteins and cleaning compound residues	Stainless-steel surfaces after CIP	SPE-RP-HPLC	Sensitive, reproducible quantification for CIP validation	Residual milk proteins and cleaning compound residues
[182]	Multiple QACs (e.g., BACs, DDACs)	Milk, yogurt, powdered dairy products	UHPLC-MS/MS	Low µg/kg detection, supports routine regulatory monitoring	Multiple QACs (e.g., BACs, DDACs)
[183]	N-(3-aminopropyl)-N-dodecylpropane-1,3-diamine	Dairy matrices	LC-MS/MS	High selectivity and sensitivity for non-quaternary biocides	N-(3-aminopropyl)-N-dodecylpropane-1,3-diamine
[184]	BACs and DDACs	Milk and dairy products	Multiresidue LC-MS/MS	EU-criteria-compliant performance for official control	BACs and DDACs
[185]	>100 biocidal actives: QACs, isothiazolinones, phenolics, etc.	Dairy products and slurry feed	LC-HRMS after acetate-buffered QuEChERS	LOQs 10 ng/g, full SANTE compliance, wide screening capability	>100 biocidal actives: QACs, isothiazolinones, phenolics, etc.
[186]	Representative disinfectant actives	Wash-water, wastewater, sanitation fluids (in-field)	Portable miniaturized LC	Low-solvent, rapid on-site screening, lower sensitivity than benchtop LC	Representative disinfectant actives
[187]	Quaternary and non-quaternary cationic substances	Cleaning-validation samples	Electromembrane extraction and RP-HPLC	Enhanced selectivity and sensitivity for trace residues	Quaternary and non-quaternary cationic substances
[188]	Multiple polar/semi-polar analytes	Surface and rinse-water samples	LC with modified stationary phases	Improved selectivity using poly(dimethyldiphenylsiloxane)-nitrile phases	Multiple polar/semi-polar analytes
[189]	Multiple disinfectant/antiseptic residues	Cleaning and sanitation samples	Magnetic SPE (functionalized nanoparticles) and HPLC–MS/MS	Efficient enrichment, reduced preparation time, high recovery	Multiple disinfectant/antiseptic residues

https://doi.org/10.3390/separations13010039

Table 7. *Cont.*

Ref.	Class/Chemical Group	Target Analytes	Matrix	Analytical Technique(s)	Key Analytical Notes
Cleaning and disinfectant agent residues					
[190]	Residues recovered from stainless-steel sampling	Stainless-steel surfaces (CIP validation)	Recovery testing and HPLC-based determination	Multi-step reconditioning improves reproducibility (>90% recovery)	Residues recovered from stainless-steel sampling
Lubricant and technical fluid contamination					
[194]	Lubricant-derived migrants from food contact materials	Lipid-like lubricant fractions	Food contact simulants, packaging systems	Normal-phase LC	Demonstrates migration of lubricant components into simulants, supports inclusion in routine monitoring.
[195]	Biodegradable lubricants	Polar degradation by-products	Ester-based biodegradable lubricants, residues in processing environments	LC-based chromatographic monitoring	Detects hydrolysis-derived low-MW compounds not captured by conventional residue methods.
[196]	Natural oil–based lubricants	Oxidation and transformation products	Lubricant residues in industrial settings	LC/MS	Monitors oxidative degradation pathways, highlights need for dedicated methods.
[197]	Mineral oil hydrocarbons (MOSH)	MOSH fractions	Foods exposed during harvesting, transport, storage	On-line HPLC-GC-FID	Shows transfer of MOSH to food, high relevance for fat-rich matrices.
[198]	Mineral oil hydrocarbons (MOSH)	MOSH fractions in edible oils	Edible oils, industrial fluids	Large-scale SPE and GC-FID	Enhances sensitivity and robustness for MOSH profiling in oils.
[199]	Lubricant additives and residues	Multi-class lubricant-related compounds	Incidental contamination events, equipment surfaces	Combined GC-MS/MS and LC-MS/MS	Supports source differentiation and profiling of additive signatures.
[200]	Ester-based lubricants	Oxidation, hydrolysis and tribological degradation products	Lubricants aged under thermal/ tribological stress	LC-ESI-MS	Distinguishes primary lubricant constituents from degradation by-products at trace levels.
[201]	Mineral oil hydrocarbons (MOSH/MOAH)	MOSH and MOAH fractions	Complex food and food contact matrices	Standardized LC–GC approaches	Enables selective determination, aligned with harmonized monitoring criteria.
[202]	Antioxidants and degradation markers	Lubricant-derived antioxidants, oxidative markers	Food contact materials and technical fluids	TLC and spray-MS	Provides rapid screening capability for lubricant-related contaminants.
Pesticide Residues from Indoor Pest Control					
[210]	Anticoagulant rodenticides	First- and second-generation anticoagulants	Contaminated food materials	HPLC-DAD, HPLC-fluorescence	Qualitative multi-residue screening in complex matrices
[211]	Anticoagulant rodenticides	Multiple rodenticides	Accidentally contaminated foods	Rapid LC-MS/MS	Short run times, reduced sample handling for routine control
[212]	Anticoagulant rodenticides (chiral)	Bromadiolone, difenacoum, brodifacoum stereoisomers	Biological matrices, contamination verification	Chiral LC-MS/MS	Enantiomer-specific quantification, differential persistence analysis

https://doi.org/10.3390/separations13010039

Table 7. *Cont.*

Ref.	Class/Chemical Group	Target Analytes	Matrix	Analytical Technique(s)	Key Analytical Notes
Pesticide Residues from Indoor Pest Control					
[213]	Neonicotinoids	Imidacloprid, acetamiprid, clothianidin and metabolites	Wine matrices	Direct-injection LC-MS/MS	Simultaneous multi-residue determination with minimal prep
[214]	Neonicotinoids	Imidacloprid	Leafy vegetables	Micellar-based microextraction and HPLC	Low-solvent extraction, selective imidacloprid determination
[215]	Organophosphates	Dimethoate, quinalphos, chlorpyrifos	Bean vegetables	Modified QuEChERS and GC-FTD	Quantitative screening for high-consumption crops
[216]	Organophosphates	Chlorpyrifos, trichlorfon	Aquaculture products (shrimp)	LC-MS/MS, GC-MS/MS	Low μg/kg detection, monitoring of illegal/ off-label use
[217]	Organophosphates and pyrethroids	Malathion, λ-cyhalothrin	Zucchini	GC-NPD, GC-ECD	Optimized solvent extraction, dual-detector confirmation
[218]	Multiple insecticide classes	Fipronil, indoxacarb, avermectin, pyridaben	Beverages (tea, juices)	d-SPME and DLLME and HPLC-MS/MS	Trace-level (ng/L) preconcentration with minimal solvent
[219]	Neonicotinoids (multi-residue)	Multiple neonicotinoids	Fruit samples	Magnetic SPE (Ti-modified silica-PSA) and UPLC-HRMS	High clean-up efficiency, improved selectivity vs. conventional SPE
[220]	Various insecticides	Emamectin benzoate, spirotetramat, tolfenpyrad, fipronil	Green and red chilli	LC-MS/MS	Evaluation of washing/ processing effects, dehydration concentrates residues

5. From Standards to Science: Integrating ISO 22002-100:2025 with GFSI Requirements Through Chromatographic Evidence

The publication of ISO 22002-100:2025 has strengthened the alignment between prerequisite programme requirements across the food, feed, and packaging sectors and the core expectations embedded in internationally recognized food safety standards, including those defined in major GFSI-recognized certification schemes. When compared with the detailed provisions of FSSC 22000, BRCGS Food Safety, and IFS Food, a consistent convergence emerges across key aspects that directly affect product integrity and certification performance, including allergen control, food fraud mitigation, traceability, and the management of chemical and packaging-related contaminants. ISO 22002-100:2025 establishes explicit operational prerequisites for validated allergen-cleaning procedures, structured vulnerability assessments for authenticity, traceable control of externally provided materials, and analytical verification of chemical hazard controls, including residues arising from cleaning chemicals, disinfectants, lubricants, and packaging migrants [1]. These requirements are closely aligned with those found in FSSC, BRCGS, and IFS, all of which emphasize demonstrable control measures, documented verification of PRPs, and the use of scientifically justified methods for assessing the effectiveness of allergen, authenticity, and chemical contamination controls [221–223].

The increasing consistency among these standards reflects the broader effort of ISO to ensure that its technical requirements remain compatible with globally accepted expectations for food safety assurance, thereby supporting uniformity in compliance interpretation, auditing depth, and certification reliability. In this context, chromatographic analysis emerges as a natural technical interface between ISO- and GFSI-recognized requirements,

https://doi.org/10.3390/separations13010039

providing objective and reproducible measurements that allow organizations to demonstrate conformity.

Building on this alignment, ISO 22002-100:2025 introduces a decisive conceptual advance by redefining prerequisite programmes as measurable performance obligations rather than procedural requirements. The standard explicitly expects organizations to demonstrate the actual effectiveness of their preventive controls through quantifiable outcomes, moving PRP verification from documentation-based confirmation to scientifically supported assessment [1]. This expectation is consistent with the requirements of FSSC, BRCGS, and IFS, all of which mandate that verification activities be based on technically adequate, scientifically justified methods for determining whether allergen control, authenticity protection, traceability measures, and chemical contamination controls are functioning as intended [221–223].

In a risk-based compliance model, ISO 22002-100:2025 integrates PRP specifications with chemical food safety requirements by establishing the need for verification evidence that controls operate as intended. In Europe, chemical hazards relevant to PRPs, such as residues from cleaning chemicals and lubricants, pest control chemicals, and packaging migrants, are managed through documented controls supported by measurable verification. In this context, chromatographic results become compliance-relevant evidence by substantiating control effectiveness against predefined acceptance criteria and enabling audit-ready demonstration that PRPs deliver measurable performance rather than procedural conformity alone. Here, "chemical food safety" within the FSMS is understood as risk-based control of chemical hazards arising from incoming materials and packaging interactions, processing, hygiene operations, and equipment maintenance [1,3]. Operationally, this approach aligns hazard identification and PRP control measures with defined verification activities, documented acceptance criteria/decision limits, and corrective actions when criteria are not met [3,4].

From a PRP verification perspective, method selection is fit-for-purpose and risk-based: high-throughput screening may support routine monitoring, whereas confirmatory chromatographic evidence is required for specificity, traceability, or results suitable for regulatory use. Practical implementation is determined by matrix effects, sampling design, sample preparation requirements, instrument availability, and the feasibility of validated performance characteristics (e.g., selectivity, sensitivity, and reproducibility) across sites. Accordingly, verification is often structured as a two-step scheme in which predefined performance criteria (acceptance criteria/decision limits) guide the transition from routine checks to targeted LC-MS/MS or GC-MS confirmatory analysis, generating traceable, measurable evidence suitable for risk-based verification and auditing [3,4].

Within this context, the "From Standards to Science" transition becomes evident: analytical data are no longer supplementary but constitute the primary means of demonstrating control effectiveness. Chromatographic measurements in particular provide the level of precision, reproducibility, and traceability needed to substantiate PRP performance, thereby meeting both ISO's strengthened technical expectations and the evidence-based verification approach required by GFSI-recognized certification schemes [1,221–223].

As demonstrated across the allergen, authenticity, and chemical contaminant applications reviewed in Sections 2–4, chromatographic techniques constitute the primary analytical basis through which organizations can generate the objective evidence required to demonstrate effective PRP performance. Advanced analytical methods such as LC-MS/MS, GC-MS/MS, and LC-HRMS provide the quantitative precision, structural specificity, and traceability needed to verify whether allergen removal, packaging migration controls, auxiliary chemical residues, and lubricant or pesticide contamination are maintained within acceptable limits. These data-driven outputs support the risk-based auditing approach

https://doi.org/10.3390/separations13010039

defined in ISO 22003-1:2022 [4], enabling auditors and operators to assess control effectiveness using measurable and reproducible criteria rather than procedural descriptions alone. Accordingly, chromatographic evidence has evolved from a supplementary analytical tool to a compliance-critical component of modern food safety verification.

Despite this progress, notable analytical gaps remain: many chromatographic methods for cleaning agents and disinfectants have been validated primarily in water, surface swabs, or CIP-rinse matrices rather than in complex food matrices, and methods for lubricant-derived contaminants remain largely focused on MOSH/MOAH fractions in oils and packaging materials rather than diverse food products. These limitations underline the need for broader method development, matrix-appropriate extraction protocols, and expanded reference materials to fully support the evidence-based verification expectations embedded in ISO- and GFSI-recognized requirements.

In conclusion, the increasing alignment between ISO- and GFSI-recognized requirements creates a compliance environment in which PRPs must be supported by measurable verification rather than procedural descriptions. As auditing becomes more evidence-driven, analytical data, particularly chromatographic measurements, provide the technical basis for demonstrating control of allergens, authenticity risks, chemical residues, and packaging-related contaminants. In this context, analytical verification is no longer optional but an essential component of credible and reproducible certification outcomes. The integration of chromatographic evidence into PRP verification therefore represents a necessary condition for consistent, technically defensible food safety assurance across the global food chain.

6. Conclusions

The continuous updating of international food safety standards, particularly the transition toward more technically explicit PRP requirements, underscores the increasing need for verification approaches that are both analytically robust and adaptable to emerging compliance requirements.

Chromatographic analysis has become an essential component of current food safety verification, supporting the transition to measurable, evidence-based control across allergens, authenticity risks, chemical contaminants, and packaging-related migrants within ISO 22002-100:2025-aligned PRP verification. As regulatory requirements and ISO-GFSI alignment increasingly emphasize demonstrable PRP performance, analytical data provide the most reliable basis for validating preventive measures and assessing system robustness.

From a compliance perspective, chromatographic verification adds the greatest value when it delivers targeted, matrix-appropriate evidence for PRP effectiveness (e.g., allergen-cleaning verification, authenticity-related markers, and control of chemical/packaging-related contaminants). This is strengthened when results are interpreted against clearly defined acceptance criteria and decision limits and documented in a traceable form suitable for risk-based verification and auditing. Practical implementation therefore benefits from verification designs that distinguish routine monitoring from confirmatory testing when higher specificity is required. Several analytical approaches reported in the literature do not yet consistently meet the level of matrix-specific validation, decision limit establishment, and performance harmonization needed to support PRP verification under ISO 22002-100:2025, meaning that analytical capability does not always translate into audit-ready compliance documentation. In addition, many published chromatographic methods were developed primarily for research studies or regulatory monitoring rather than for PRP verification within management system standards, which can limit their direct transferability to audit-driven, risk-based compliance.

https://doi.org/10.3390/separations13010039

Ongoing developments in liquid and gas chromatographic techniques, together with expanded availability of reference materials and more consistent validation practices, will enhance the scientific robustness of food safety compliance. These advancements also indicate the need for closer coordination among laboratories, operators, and certification bodies to ensure the consistent and reliable application of analytical verification in food safety decision-making.

In this context, routine analytical verification is essential for demonstrating that PRPs operate as intended, particularly when chromatographic results are interpreted against pre-defined acceptance criteria/decision limits and documented in a traceable, audit-ready form suitable for risk-based verification and auditing. By focusing on PRP verification and auditability, this review distinguishes analytical innovation from compliance readiness and highlights method development priorities relevant to ISO- and GFSI-aligned systems. Future research should prioritize developing matrix-appropriate methods and expanding analytical coverage for emerging contaminants, ensuring that verification practices keep pace with evolving PRP requirements.

Author Contributions: Conceptualization, E.G.K. and V.F.S.; data curation, E.G.K. and N.A.F.N.; writing—original draft, E.G.K. and N.A.F.N.; writing—review and editing, E.G.K. and V.F.S.; supervision, V.F.S. All authors have read and agreed to the published version of the manuscript.

Funding: This research received no external funding.

Data Availability Statement: No new data were created or analyzed in this study.

Conflicts of Interest: Author Eftychia G. Karageorgou is employed by the company NBIS P.C. Cargo Inspections and ISO Certification. This work was conducted independently of her professional role. All authors declare that the research was conducted in the absence of any commercial or financial relationships that could be construed as a potential conflict of interest.

Abbreviations

The following abbreviations are used in this manuscript:

PRPs	Prerequisite programmes
NIAS	Non-intentionally added substances
HPLC	High-performance liquid chromatography
GC	Gas chromatography
GFSI	Global Food Safety Initiative
ED05	Eliciting Dose for 5% of the allergic population
VITAL	Voluntary Incidental Trace Allergen Labelling
FAO	Food and Agriculture Organization
WHO	World Health Organization
RfDs	Health-based reference doses
ELISAs	Enzyme-linked immunosorbent assay
LFDs	Lateral flow devices
PCR	Polymerase chain reaction
LC-MS/MS	Liquid chromatography–tandem mass spectrometry
MRM	Multiple reaction monitoring
HRMS	High-resolution mass spectrometry
DIA	Data-independent acquisition
ESI	Ion-exchange chromatography with electrospray ionization
MALDI	Matrix-assisted laser desorption/ionization

https://doi.org/10.3390/separations13010039

SRM	Selected reaction monitoring
QC	Quality Control
AOAC	Association of Official Analytical Collaboration
CEN/TC 275/WG12	European Committee for Standardization, Technical Committee 275—Food Analysis—Working Group 12: Food Allergens
iFAAM	Integrated Approaches to Food Allergen and Allergy Risk Management
CRMs	Certified reference materials
SPE	Solid-phase extraction
TLC	Thin-layer chromatography
IEC	Ion-exchange chromatography
DAS	Double Antibody Sandwich
LFIA	Lateral flow immunoassay
LOQ	Limit of Quantitation
LOD	Limit of Detection
UHPLC	Ultra-high-performance liquid chromatography
VACCP	Vulnerability Assessment and Critical Control Points
TACCP	Threat Assessment and Critical Control Points
VOC	Volatile organic compounds
HPTLC-MS	High-performance thin-layer chromatography coupled with mass spectrometry
EVOO	Extra virgin olive oil
LC-IRMS	Liquid chromatography–isotope ratio mass spectrometry
NMR	Nuclear Magnetic Resonance
PDO	Protected Designation of Origin
PGI	Protected Geographical Indication
CSIA	Compound-specific isotope analysis
IRMS	Isotope ratio mass spectrometry
MIR	Mid-Infrared
NIR	Near-Infrared
HS-SPME-GC-MS	Headspace solid-phase microextraction–gas chromatography–mass spectrometry
GC-C-IRMS	Gas chromatography–combustion–isotope ratio mass spectrometry
TGA	Thermogravimetric Analysis
TAG	Triacylglycerols
DAG	Diacylglycerols
UV	Ultraviolet
SFC-Q-TOF-MS	Supercritical fluid chromatography–quadrupole time-of-flight-mass spectrometry
RP-HPLC	Reversed-phase high-performance liquid chromatography
MALDI-TOF-MS	Matrix-assisted laser desorption/ionization–time-of-flight–mass spectrometry
UHPLC-ELSD/UV	Ultra-high-performance liquid chromatography–evaporative light scattering detection/ultraviolet detection
PCA	Principal Component Analysis
PLS-DA	Partial Least-Squares Discriminant Analysis
HMF	Hydroxymethylfurfural
HPLC-RID	High-performance liquid chromatography-refractive index detection
GC-FID	Gas chromatography-flame ionization detection
GC-IMS	GC-ion mobility spectrometry
HPLC-DAD	High-performance liquid chromatography with diode array detection
ICP-OES	Inductively Coupled Plasma—Optical Emission Spectrometry
ID	Identity document
FT-NIR	Fourier-transform NIR spectroscopy
MGO	Methylglyoxal

https://doi.org/10.3390/separations13010039

OPLS-DA	Orthogonal Partial Least Squares-Discriminant Analysis
GC-GC/TOF-MS	Comprehensive two-dimensional gas chromatography coupled with time-of-flight mass spectrometry
DI-SPME-GC-MS	Direct immersion–solid-phase microextraction–gas chromatography–mass spectrometry
MOSHs	Mineral oil saturated hydrocarbons
MOAHs	Mineral oil aromatic hydrocarbons
SVOCs	Semi-volatile organic compounds
BPA	Bisphenol A
BPS	Bisphenol S
MOHs	Mineral oil hydrocarbons
CIP	Cleaning-in-place
SPE-RP-HPLC	Solid-phase extraction combined with reversed-phase high-performance liquid chromatography
QACs	Quaternary ammonium compounds
BACs	Benzalkonium chlorides
DDACs	Dialkyl-dimethylammonium chlorides
QuEChERS	Quick Easy Cheap Effective Rugged Safe
SANTE	Directorate-General for Health and Food Safety of the European Commission
GC-FTD	Gas chromatography–flame thermionic detector
GC-NPD	Gas chromatography with nitrogen-phosphorus detection
GC-ECD	Gas chromatography–electron capture detector
DLLME	Dispersive liquid–liquid microextraction
PSA	Porous Silica
FSSC	Food Safety System Certification 22000
BRCGS	British Retail Consortium Global Standards
IFS	International Featured Standards

References

1. *ISO 22002-100:2025*; Prerequisite Programmes on Food Safety—Part 100: Food, Feed and Packaging Supply Chain, 1st ed. ISO: Geneva, Switzerland, 2025.
2. *ISO/TS 22002-7:2025*; Prerequisite Programmes on Food Safety—Part 7: Retail and Wholesale, 1st ed. ISO: Geneva, Switzerland, 2025.
3. *ISO 22000:2018*; Food Safety Management Systems—Requirements for Any Organization in the Food Chain, 1st ed. ISO: Geneva, Switzerland, 2018.
4. *ISO 22003-1:2022*; Food Safety—Part 1: Requirements for Bodies Providing Audit and Certification of Food Safety Management Systems, 1st ed. ISO: Geneva, Switzerland, 2022.
5. Karageorgou, E.G.; Tzikas, Z. Regulatory Requirements as Control Measures: A HACCP-Based Model for Food Safety Compliance in the EU. *J. Assoc. Greek Chem.* **2025**, 2, 44–51. [CrossRef]
6. European Parliament and Council. *Regulation (EU) No 1169/2011 of the European Parliament and of the Council of 25 October 2011 on the Provision of Food Information to Consumers*; Annex II; EU: Brussels, Belgium, 2011.
7. Millard, P.; Paine, S.; O'Hagan, S.; Hipkiss, J. Traceability of Allergenic Foods in the Food Chain. In *Handbook of Food Allergen Detection and Control*, 1st ed.; Woodhead Publishing: Cambridge, UK, 2025; pp. 19–40. [CrossRef]
8. Crevel, R. Developing Food Allergen Management Thresholds. In *Handbook of Food Allergen Detection and Control*, 2nd ed.; Woodhead Publishing: Cambridge, UK, 2025; pp. 51–73.
9. FAO; WHO. *Risk Assessment of Food Allergens—Part 2: Review and Establish Threshold Levels in Foods for the Priority Allergens*; Meeting Report, Food Safety and Quality Series No. 15; FAO/WHO: Rome, Italy, 2022.
10. Lacorn, M.; Immer, U. Enzyme-Linked Immunosorbent Assays (ELISAs) for Detecting Allergens in Food. In *Handbook of Food Allergen Detection and Control*, 2nd ed.; Woodhead Publishing: Cambridge, UK, 2025; pp. 173–189. [CrossRef]
11. Baumert, J.L.; Jayasena, S. Lateral Flow Devices for Detecting Allergens in Food. In *Handbook of Food Allergen Detection and Control*, 2nd ed.; Woodhead Publishing: Cambridge, UK, 2025; pp. 191–202. [CrossRef]
12. Chen, Y.; Wang, P.; Wang, W.; Han, J.X. Optical Thin-Film Biochips for Detecting Allergens in Food. In *Handbook of Food Allergen Detection and Control*, 2nd ed.; Woodhead Publishing: Cambridge, UK, 2025; pp. 203–210. [CrossRef]

https://doi.org/10.3390/separations13010039

13. Holzhauser, T.; Röder, M. Polymerase Chain Reaction (PCR) Methods for Detecting Allergens in Foods. In *Handbook of Food Allergen Detection and Control*, 2nd ed.; Woodhead Publishing: Cambridge, UK, 2025; pp. 211–227. [CrossRef]
14. Hamilton, R.G. IgE Antibody-Based Analysis for Detecting Allergens in Food. In *Handbook of Food Allergen Detection and Control*, 2nd ed.; Woodhead Publishing: Cambridge, UK, 2025; pp. 229–249. [CrossRef]
15. Johnson, P.; Downs, M.L. Mass Spectrometry for Analysis of Food Allergens. In *Handbook of Food Allergen Detection and Control*, 2nd ed.; Woodhead Publishing: Cambridge, UK, 2025; pp. 415–435.
16. Monaci, L.; De Angelis, E.; Montemurro, N.; Pilolli, R. Comprehensive Overview and Recent Advances in Proteomics MS Based Methods for Food Allergens Analysis. *TrAC Trends Anal. Chem.* **2018**, *106*, 21–36. [CrossRef]
17. Fæste, C.K.; Rønning, H.T.; Christians, U.; Granum, P.E. Liquid Chromatography and Mass Spectrometry in Food Allergen Detection. *J. Food Prot.* **2011**, *74*, 316–345. [CrossRef]
18. Shin, J.H.; Reddy, Y.V.M.; Park, T.J.; Park, J.P. Recent Advances in Analytical Strategies and Microsystems for Food Allergen Detection. *Food Chem.* **2022**, *371*, 131120. [CrossRef] [PubMed]
19. Zhang, Y.; Lei, S.; Zou, W.; Wang, L.; Yan, J.; Zhang, X.; Zhang, W.; Yang, Q. Research Progress on Detection Methods for Food Allergens. *J. Food Compos. Anal.* **2025**, *137*, 106906. [CrossRef]
20. Croote, D.; Quake, S.R. Food Allergen Detection by Mass Spectrometry: The Role of Systems Biology. *npj Syst. Biol. Appl.* **2016**, *2*, 16022. [CrossRef]
21. Hu, W.; Zhang, X.; Shen, Y.; Meng, X.; Wu, Y.; Tong, P.; Li, X.; Chen, H.; Gao, J. Quantifying Allergenic Proteins Using Antibody-Based Methods or Liquid Chromatography–Mass Spectrometry/Mass Spectrometry: A Review About the Influence of Food Matrix, Extraction, and Sample Preparation. *Compr. Rev. Food Sci. Saf.* **2024**, *23*, e70029. [CrossRef] [PubMed]
22. Linacero, R.; Cuadrado, C. New Research in Food Allergen Detection. *Foods* **2022**, *11*, 1520. [CrossRef]
23. Flanagan, S.; Hirst, B.; Miguel, S.J. Validation, Standardization, and Harmonization of Analytical Methods and Test Kits for Detecting Allergens in Food. In *Handbook of Food Allergen Detection and Control*, 2nd ed.; Woodhead Publishing: Cambridge, UK, 2025; pp. 251–264. [CrossRef]
24. Holcombe, G.; Walker, M.J.; Singh, M.; Gray, K.; Cowen, S.; Ellison, S.L.R.; Rogers, A.; Balasundaram, A.; Burns, M.; Mills, E.N.C. Clinically and Industrially Relevant Incurred Reference Materials to Improve Analysis of Food Allergens, Milk, Egg, Almond, Hazelnut and Walnut. *Food Chem.* **2024**, *434*, 137391. [CrossRef]
25. Downs, M.L. Milk as a Food Allergen. In *Handbook of Food Allergen Detection and Control*, 2nd ed.; Woodhead Publishing: Cambridge, UK, 2025; pp. 301–316.
26. Qi, K.; Liu, T.; Yang, Y.; Zhang, J.; Yin, J.; Ding, X.; Qin, W.; Yang, Y. A Rapid Immobilized Trypsin Digestion Combined with Liquid Chromatography–Tandem Mass Spectrometry for the Detection of Milk Allergens in Baked Food. *Food Control* **2019**, *102*, 179–187. [CrossRef]
27. Bianco, M.; Calvano, C.D.; Ventura, G.; Losito, I.; Cataldi, T.R.I. Determination of Hidden Milk Allergens in Meat-Based Foodstuffs by Liquid Chromatography Coupled to Electrospray Ionization and High-Resolution Tandem Mass Spectrometry. *Food Control* **2022**, *131*, 108443. [CrossRef]
28. Ji, J.; Zhu, P.; Pi, F.; Sun, C.; Sun, J.; Jia, M.; Ying, C.; Zhang, Y.; Sun, X. Development of a Liquid Chromatography-Tandem Mass Spectrometry Method for Simultaneous Detection of the Main Milk Allergens. *Food Control* **2017**, *74*, 79–88. [CrossRef]
29. Cucu, T.; De Muer, N.; Trashin, S.; De Meulenaer, B. Mechanisms Behind Matrix–Protein Interactions Influencing Receptor-Based and Chromatographic Detection of Food Allergens: A Case Study with a Fruit Based Snack. *Food Control* **2015**, *47*, 641–646. [CrossRef]
30. Lamberti, C.; Acquadro, E.; Corpillo, D.; Giribaldi, M.; Decastelli, L.; Garino, C.; Arlorio, M.; Ricciardi, C.; Cavallarin, L.; Giuffrida, M.G. Validation of a Mass Spectrometry-Based Method for Milk Traces Detection in Baked Food. *Food Chem.* **2016**, *199*, 119–127. [CrossRef]
31. Lock, S. Gluten Detection and Speciation by Liquid Chromatography Mass Spectrometry (LC-MS/MS). *Foods* **2014**, *3*, 13–29. [CrossRef] [PubMed]
32. Scherf, K.A.; Poms, R.E. Recent Developments in Analytical Methods for Tracing Gluten. *J. Cereal Sci.* **2016**, *67*, 112–122. [CrossRef]
33. Martinez-Aviño, A.; Moliner-Martinez, Y.; Molins-Legua, C.; Campins-Falcó, P. Colorimetric Analysis Platform Based on Thin Layer Chromatography for Monitoring Gluten Cross-Contamination in Food Industry. *Food Chem.* **2024**, *448*, 139025. [CrossRef]
34. Zeng, J.; Ma, F.; Zhai, L.; Du, C.; Zhao, J.; Li, Z.; Wang, J. Recent Advance in Sesame Allergens: Influence of Food Processing and Their Detection Methods. *Food Chem.* **2024**, *448*, 139058. [CrossRef]
35. Ma, X.; Li, H.; Zhang, J.; Huang, W.; Han, J.; Ge, Y.; Sun, J.; Chen, Y. Comprehensive Quantification of Sesame Allergens in Processed Food Using Liquid Chromatography-Tandem Mass Spectrometry. *Food Control* **2020**, *107*, 106744. [CrossRef]
36. Li, J.; Kang, W.; Zhang, J.; Ge, Y.; Yu, N.; Chen, Y. Selection of Signature Peptide Biomarkers for the Sesame Allergens in Commercial Food Based on LC-MS/MS. *Food Chem.* **2025**, *463*, 141392. [CrossRef]
37. Taylor, S.L.; Downs, M.L.; Baumert, J.L. Soybeans as a Food Allergen. In *Handbook of Food Allergen Detection and Control*, 2nd ed.; Woodhead Publishing: Cambridge, UK, 2025; pp. 343–368. [CrossRef]

https://doi.org/10.3390/separations13010039

38. Chen, S.; Yang, C.; Downs, M. Targeted Mass Spectrometry Quantification of Total Soy Protein Residues from Commercially Processed Ingredients for Food Allergen Management. *J. Proteom.* **2021**, *239*, 104194. [CrossRef]
39. Krager, J.; Baumert, J.L.; Downs, M.L. Quantification of Soy-Derived Ingredients in Model Bread and Frankfurter Matrices with an Optimized Liquid Chromatography–Tandem Mass Spectrometry External Standard Calibration Workflow. *J. Food Prot.* **2022**, *85*, 311–322. [CrossRef]
40. Karimi, Z.; Campbell, K.; Kevei, Z.; Patriarca, A.; Koidis, A.; Anastasiadi, M. A Critical Review of Conventional and Emerging Technologies for the Detection of Contaminants, Allergens and Adulterants in Plant-Based Milk Alternatives. *Curr. Res. Food Sci.* **2025**, *10*, 101067. [CrossRef]
41. Sakai, S.; Adachi, R.; Teshima, R. Detection and Control of Eggs as a Food Allergen. In *Handbook of Food Allergen Detection and Control*; Mills, E.N.C., Wichers, H.J., Eds.; Woodhead Publishing: Cambridge, UK, 2015; pp. 313–340. [CrossRef]
42. Ma, X.; Liang, R.; Yang, X.; Gou, J.; Li, Y.; Lozano-Ojalvo, D. Simultaneous Separation of the Four Major Allergens of Hen Egg White. *J. Chromatogr. B* **2020**, *1152*, 122231. [CrossRef] [PubMed]
43. Kiyota, K.; Yoshimitsu, M.; Uchida, K.; Kajimura, K. Development of a Liquid Chromatography–Tandem Mass Spectrometry Method for Simultaneous Quantification of Hen's Egg White Allergens Gal d 1–4 in Fresh and Processed Eggs. *Food Chem.* **2021**, *345*, 128022. [CrossRef] [PubMed]
44. Yang, S.; Chen, J.; Abdallah, M.F.; Lin, H.; Yang, P.; Li, J.; Zhang, R.; Li, Q.; Lu, P.; Liu, S.; et al. An Integrated Calibration Strategy for the Development and Validation of an LC-MS/MS Method for Accurate Quantification of Egg Allergens (Gal d 1–6) in Foods. *Food Chem.* **2024**, *438*, 137922. [CrossRef] [PubMed]
45. Wong, G.W.K. Detection and Control of Fish, Shellfish, and Mollusks as Food Allergens. In *Handbook of Food Allergen Detection and Control*, 2nd ed.; Mills, E.N.C., Wichers, H.J., Eds.; Woodhead Publishing: Cambridge, UK, 2025; pp. 389–398.
46. Sun, L.; Lin, H.; Li, Z.; Sun, W.; Wang, J.; Wu, H.; Ge, M.; Ahmed, I.; Pavase, T.R. Development of a Method for the Quantification of Fish Major Allergen Parvalbumin in Food Matrix via Liquid Chromatography–Tandem Mass Spectrometry with Multiple Reaction Monitoring. *Food Chem.* **2019**, *276*, 358–365. [CrossRef]
47. Posada-Ayala, M.; Alvarez-Llamas, G.; Maroto, A.S.; Maes, X.; Muñoz-Garcia, E.; Villalba, M.; Rodríguez, R.; Perez-Gordo, M.; Vivanco, F.; Pastor-Vargas, C.; et al. Novel Liquid Chromatography–Mass Spectrometry Method for Sensitive Determination of the Mustard Allergen Sin a 1 in Food. *Food Chem.* **2015**, *183*, 58–63. [CrossRef]
48. Chen, Y.; Wu, Y.J.; Deng, T.T. Detection and Control of Mustard and Sesame as Food Allergens. In *Handbook of Food Allergen Detection and Control*, 1st ed.; Mills, E.N.C., Wichers, H.J., Eds.; Woodhead Publishing: Cambridge, UK, 2015; pp. 391–408.
49. Yang, S.; Sun, F.; Wu, X.; Huang, P.; Qu, J.; He, R.; Cui, Y.; Xi, J. Establishment of DAS-ELISA and Lateral Flow Immunochromatography for the Detection of Peanut Allergen Ara h 3 after Heat-Moisture Treatment. *Food Chem.* **2025**, *472*, 142910. [CrossRef]
50. Downs, M.L.; Baumert, J.L.; Taylor, S.L.; Mills, E.N.C. Mass Spectrometric Analysis of Allergens in Roasted Walnuts. *J. Proteom.* **2016**, *142*, 62–69. [CrossRef]
51. Fan, S.; Ma, J.; Liu, Z.; Ning, Y.; Cao, M.; Li, Q.; Zhang, Y. Determination of Egg and Milk Allergen in Food Products by Liquid Chromatography-Tandem Mass Spectrometry Based on Signature Peptides and Isotope-Labeled Internal Standard. *Food Sci. Hum. Wellness* **2023**, *12*, 728–736. [CrossRef]
52. Torii, A.; Seki, Y.; Arimoto, C.; Hojo, N.; Iijima, K.; Nakamura, K.; Ito, R.; Yamakawa, H.; Akiyama, H. Development of a Simple and Reliable LC-MS/MS Method to Simultaneously Detect Walnut and Almond as Specified in Food Allergen Labelling Regulations in Processed Foods. *Curr. Res. Food Sci.* **2023**, *6*, 100444. [CrossRef]
53. Parker, C.H.; Khuda, S.E.; Pereira, M.; Ross, M.M.; Fu, T.-J.; Fan, X.; Wu, Y.; Williams, K.M.; DeVries, J.; Pulvermacher, B.; et al. Multi-Allergen Quantitation and the Impact of Thermal Treatment in Industry-Processed Baked Goods by ELISA and Liquid Chromatography-Tandem Mass Spectrometry. *J. Agric. Food Chem.* **2015**, *63*, 10669–10680. [CrossRef]
54. Boo, C.C.; Parker, C.H.; Jackson, L.S. A Targeted LC-MS/MS Method for the Simultaneous Detection and Quantitation of Egg, Milk, and Peanut Allergens in Sugar Cookies. *J. AOAC Int.* **2018**, *101*, 108–117. [CrossRef]
55. Xiong, W.; Parker, C.H.; Boo, C.C.; Fiedler, K.L. Comparison of Allergen Quantification Strategies for Egg, Milk, and Peanut in Food Using Targeted LC–MS/MS. *Anal. Bioanal. Chem.* **2021**, *413*, 5755–5766. [CrossRef]
56. Monaci, L.; Pilolli, R.; De Angelis, E.; Godula, M.; Visconti, A. Multi-Allergen Detection in Food by Micro High-Performance Liquid Chromatography Coupled to a Dual Cell Linear Ion Trap Mass Spectrometry. *J. Chromatogr. A* **2014**, *1358*, 136–144. [CrossRef] [PubMed]
57. Gomaa, A.; Boye, J. Simultaneous Detection of Multi-Allergens in an Incurred Food Matrix Using ELISA, Multiplex Flow Cytometry and Liquid Chromatography Mass Spectrometry (LC-MS). *Food Chem.* **2015**, *175*, 585–592. [CrossRef]
58. Bi, Y.; Liu, C.; Pan, F.; Peng, W.; Fang, X.; Che, H.; Tian, W. Allergenicity Risk in Animal-Based Food Proteins: Source, Dietary Factors Effect, Allergen Detection and Processing Modification Methods. *Trends Food Sci. Technol.* **2024**, *153*, 104726. [CrossRef]
59. Planque, M.; Arnould, T.; Dieu, M.; Delahaut, P.; Renard, P.; Gillard, N. Advances in Ultra-High Performance Liquid Chromatography Coupled to Tandem Mass Spectrometry for Sensitive Detection of Several Food Allergens in Complex and Processed Foodstuffs. *J. Chromatogr. A* **2016**, *1464*, 115–123. [CrossRef]

https://doi.org/10.3390/separations13010039

60. Stella, R.; Sette, G.; Moressa, A.; Gallina, A.; Aloisi, A.M.; Angeletti, R.; Biancotto, G. LC-HRMS/MS for the Simultaneous Determination of Four Allergens in Fish and Swine Food Products. *Food Chem.* **2020**, *331*, 127276. [CrossRef]
61. New, L.S.; Stahl-Zeng, J.; Schreiber, A.; Cafazzo, M.; Liu, A.; Brunelle, S.; Liu, H.-F. Detection and Quantitation of Selected Food Allergens by Liquid Chromatography with Tandem Mass Spectrometry: First Action 2017.17. *J. AOAC Int.* **2020**, *103*, 570–583. [CrossRef]
62. Gavage, M.; Van Vlierberghe, K.; Dieu, M.; Renard, P.; Arnould, T.; De Loose, M.; Gevaert, K.; Gillard, N.; Van Poucke, C. Development and Validation of a Quantitative Method for Multiple Allergen Detection in Food Using Concatemer-Based Isotope Dilution Mass Spectrometry. *J. AOAC Int.* **2022**, *105*, 1585–1595. [CrossRef]
63. Heick, J.; Fischer, M.; Pöpping, B. First Screening Method for the Simultaneous Detection of Seven Allergens by Liquid Chromatography Mass Spectrometry. *J. Chromatogr. A* **2011**, *1218*, 938–943. [CrossRef]
64. Pilolli, R.; De Angelis, E.; Monaci, L. Streamlining the Analytical Workflow for Multiplex MS/MS Allergen Detection in Processed Foods. *Food Chem.* **2017**, *221*, 1747–1753. [CrossRef] [PubMed]
65. Torii, A.; Seki, Y.; Sasano, R.; Ishida, Y.; Nakamura, K.; Ito, R.; Iwasaki, Y.; Iijima, K.; Akiyama, H. Development of a Rapid and Reliable Method to Simultaneously Detect Seven Food Allergens in Processed Foods Using LC-MS/MS. *Food Chem. X* **2024**, *23*, 101558. [CrossRef] [PubMed]
66. Planque, M.; Arnould, T.; Delahaut, P.; Renard, P.; Dieu, M.; Gillard, N. Development of a Strategy for the Quantification of Food Allergens in Several Food Products by Mass Spectrometry in a Routine Laboratory. *Food Chem.* **2019**, *274*, 35–45. [CrossRef] [PubMed]
67. European Union. Regulation (EC) No 178/2002 of the European Parliament and of the Council of 28 January 2002 laying down the general principles and requirements of food law, establishing the European Food Safety Authority and laying procedures in matters of food safety. *Off. J. Eur. Communities* **2002**, *L 31*, 1–24.
68. European Union. Regulation (EU) 2017/625 of the European Parliament and of the Council of 15 March 2017 on official controls and other official activities performed to ensure the application of food and feed law, rules on animal health and welfare, plant health and plant protection products. *Off. J. Eur. Union* **2017**, *L 95*, 1–142.
69. Munekata, P.E.S.; Domínguez, R.; Pateiro, M.; Lorenzo, J.M. Introduction to food fraud. In *Food Toxicology and Forensics*; Galanakis, C.M., Ed.; Academic Press: Cambridge, MA, USA, 2021; pp. 1–30. [CrossRef]
70. Kennedy, S.P. History of food fraud and development of mitigation requirements and standards. In *Food Fraud: A Global Threat with Public Health and Economic Consequences*; Hellberg, R.S., Everstine, K., Sklare, S.A., Eds.; Academic Press: Cambridge, MA, USA, 2021; pp. 9–22.
71. Soon, J.M. Food fraud countermeasures and consumers: A future agenda. In *Future Foods: Global Trends, Opportunities, and Sustainability Challenges*; Academic Press: Cambridge, MA, USA, 2022; pp. 597–611. [CrossRef]
72. Rezazade, F.; Summers, J.; Ong Lai Teik, D. A Holistic Approach to Food Fraud Vulnerability Assessment. *Food Control* **2022**, *131*, 108440. [CrossRef]
73. European Commission. *Report on EU Agri-Food Fraud Suspicions (iRASFF)*; European Commission: Brussels, Belgium, 2025.
74. Everstine, K.D.; Chin, H.B.; Lopes, F.A.; Moore, J.C. Database of Food Fraud Records: Summary of Data from 1980 to 2022. *J. Food Prot.* **2024**, *87*, 100227. [CrossRef]
75. Everstine, K.; Popping, B.; Gendel, S.M. Food fraud mitigation: Strategic approaches and tools. In *Food Fraud: A Global Threat with Public Health and Economic Consequences*; Hellberg, R.S., Everstine, K., Sklare, S.A., Eds.; Academic Press: Cambridge, MA, USA, 2021; pp. 23–43.
76. Gendel, S.M. Preventing food fraud. In *Present Knowledge in Food Safety: A Risk-Based Approach Through the Food Chain*; Knowles, M.E., Anelich, L.E., Boobis, A.R., Popping, B., Eds.; Academic Press: Cambridge, MA, USA, 2023; pp. 260–265.
77. Robson, K.; Dean, M.; Haughey, S.; Elliott, C. A Comprehensive Review of Food Fraud Terminologies and Food Fraud Mitigation Guides. *Food Control* **2021**, *120*, 107516. [CrossRef]
78. Maritano, V.; Barge, P.; Biglia, A.; Comba, L.; Ricauda Aimonino, D.; Tortia, C.; Gay, P. Anticounterfeiting and Fraud Mitigation Solutions for High-Value Food Products. *J. Food Prot.* **2024**, *87*, 100251. [CrossRef]
79. Gwenzi, W.; Makuvara, Z.; Marumure, J.; Simbanegavi, T.T.; Mukonza, S.S.; Chaukura, N. Chicanery in the Food Supply Chain! Food Fraud, Mitigation, and Research Needs in Low-Income Countries. *Trends Food Sci. Technol.* **2023**, *136*, 194–223. [CrossRef]
80. Barrere, V.; Everstine, K.; Théolier, J.; Godefroy, S. Food Fraud Vulnerability Assessment: Towards a Global Consensus on Procedures to Manage and Mitigate Food Fraud. *Trends Food Sci. Technol.* **2020**, *100*, 131–137. [CrossRef]
81. Spink, J.W. The GFSI Food Fraud Prevention Compliance Development & Requirements: A Ten-Year Review. *Trends Food Sci. Technol.* **2023**, *138*, 766–773. [CrossRef]
82. Manning, L. Food defense: Types of threat, defense plans, and mitigation strategies. In *Present Knowledge in Food Safety: A Risk-Based Approach Through the Food Chain*; Knowles, M.E., Anelich, L.E., Boobis, A.R., Popping, B., Eds.; Academic Press: Cambridge, MA, USA, 2023; pp. 536–551. [CrossRef]

https://doi.org/10.3390/separations13010039

83. Butler, F.; Luijckx, N.L.; Marvin, H.J.P.; Bouzembrak, Y.; Mojtahed, V. Role of Analytical Testing for Food Fraud Risk Mitigation—A Commentary on Implementation of Analytical Fraud Testing. *Curr. Res. Food Sci.* **2021**, *4*, 301–307. [CrossRef] [PubMed]
84. Lees, M.; Reimann, L. Analytical Detection Methods and Strategies for Food Fraud. In *Food Fraud: A Global Threat with Public Health and Economic Consequences*; Hellberg, R.S., Everstine, K., Sklare, S.A., Eds.; Academic Press: Cambridge, MA, USA, 2021; pp. 45–67.
85. Marín, X.; Grau-Noguer, E.; Gervilla-Cantero, G.; Ripolles-Avila, C.; Castillo, M. Emerging Technologies for Detecting Food Fraud: A Review of the Current Landscape in the 2020s. *Trends Food Sci. Technol.* **2025**, *165*, 105313. [CrossRef]
86. Vinothkanna, A.; Dar, O.I.; Liu, Z.; Jia, A.-Q. Advanced Detection Tools in Food Fraud: A Systematic Review for Holistic and Rational Detection Method Based on Research and Patents. *Food Chem.* **2024**, *446*, 138893. [CrossRef]
87. Lei, H.; Castanheira, I.; Guan, T. Advanced Techniques Against Food Fraud. *Food Chem.* **2025**, *470*, 142832. [CrossRef]
88. Liang, C.; Xu, Z.; Liu, P.; Guo, S.; Xiao, P.; Duan, J.-A. Integrating Different Detection Techniques and Data Analysis Methods for Comprehensive Food Authenticity Verification. *Food Chem.* **2025**, *463*, 141471. [CrossRef]
89. Jiménez-Carvelo, A.M.; Pérez-Castaño, E.; González-Casado, A.; Cuadros-Rodríguez, L. One Input-Class and Two Input-Class Classifications for Differentiating Olive Oil from Other Edible Vegetable Oils by Use of the Normal-Phase Liquid Chromatography Fingerprint of the Methyl-Transesterified Fraction. *Food Chem.* **2017**, *221*, 1784–1791. [CrossRef] [PubMed]
90. Suman, M.; Cavanna, D.; Sammarco, G.; Lambertini, F.; Loffi, C. Fighting Food Frauds Exploiting Chromatography-Mass Spectrometry Technologies: Scenario Comparison Between Solutions in Scientific Literature and Real Approaches in Place in Industrial Facilities. *TrAC Trends Anal. Chem.* **2021**, *142*, 116305. [CrossRef]
91. Sharma, R.; Nath, P.C.; Lodh, B.K.; Mukherjee, J.; Mahata, N.; Gopikrishna, K.; Tiwari, O.N.; Bhunia, B. Rapid and Sensitive Approaches for Detecting Food Fraud: A Review on Prospects and Challenges. *Food Chem.* **2024**, *454*, 139817. [CrossRef]
92. Quintanilla-Casas, B.; Torres-Cobos, B.; Bro, R.; Guardiola, F.; Vichi, S.; Tres, A. The Volatile Metabolome—Gas Chromatography–Mass Spectrometry Approaches in the Context of Food Fraud. *Curr. Opin. Food Sci.* **2025**, *61*, 101235. [CrossRef]
93. Sharma, N.; Anurag; Singh, H.; Sharma, A. Application of High-Performance Thin Layer Chromatography Mass Spectrometry (HPTLC MS) in Foodomics Authenticity. *Talanta Open* **2024**, *9*, 100315. [CrossRef]
94. Lozano-Castellón, J.; Laveriano-Santos, E.P.; Abuhabib, M.M.; Pozzoli, C.; Pérez, M.; Vallverdú-Queralt, A.; Lamuela-Raventós, R.M. Proven Traceability Strategies Using Chemometrics for Organic Food Authenticity. *Trends Food Sci. Technol.* **2024**, *147*, 104430. [CrossRef]
95. Ferreira, G.; Tucker, J.; Rakola, E.; Raszap Skorbiansky, S. Fraud in Organic Foods. In *Food Fraud: A Global Threat with Public Health and Economic Consequences*; Hellberg, R.S., Everstine, K., Sklare, S.A., Eds.; Academic Press: Cambridge, MA, USA, 2021; pp. 335–350.
96. Koljančić, N.; Furdíková, K.; de Araújo Gomes, A.; Špánik, I. Wine Authentication: Current Progress and State of the Art. *Trends Food Sci. Technol.* **2024**, *150*, 104598. [CrossRef]
97. Ma, Y.; Li, Y.; Shao, F.; Lu, Y.; Meng, W.; Rogers, K.M.; Sun, D.; Wu, H.; Peng, X. Advancing Stable Isotope Analysis for Alcoholic Beverages' Authenticity: Novel Approaches in Fraud Detection and Traceability. *Foods* **2025**, *14*, 943. [CrossRef]
98. Lin, S.; Salcido-Keamo, S. Fraud in Wine and Other Alcoholic Beverages. In *Food Fraud: A Global Threat with Public Health and Economic Consequences*; Hellberg, R.S., Everstine, K., Sklare, S.A., Eds.; Academic Press: Cambridge, MA, USA, 2021; pp. 233–247.
99. Syposs, Z.; Knight, G. Fraud in Nonalcoholic Ready-to-Drink Products. In *Food Fraud: A Global Threat with Public Health and Economic Consequences*; Hellberg, R.S., Everstine, K., Sklare, S.A., Eds.; Academic Press: Cambridge, MA, USA, 2021; pp. 219–232.
100. Wadood, S.A.; Boli, G.; Xiaowen, Z.; Hussain, I.; Yimin, W. Recent Development in the Application of Analytical Techniques for the Traceability and Authenticity of Food of Plant Origin. *Microchem. J.* **2020**, *152*, 104295. [CrossRef]
101. Kennedy, S.P.; Gonzales, P.; Roungchun, J. Coffee and Tea Fraud. In *Food Fraud: A Global Threat with Public Health and Economic Consequences*; Hellberg, R.S., Everstine, K., Sklare, S.A., Eds.; Academic Press: Cambridge, MA, USA, 2021; pp. 139–150.
102. Hoffman, J.M.; Lafeuille, J.-L.; Ragupathy, S.; Newmaster, S. Spice and Herb Fraud. In *Food Fraud: A Global Threat with Public Health and Economic Consequences*; Hellberg, R.S., Everstine, K., Sklare, S.A., Eds.; Academic Press: Cambridge, MA, USA, 2021; pp. 177–218.
103. Faller, A.C.; Kesanakurti, P.; Arunachalam, T. Fraud in Grains and Cereals. In *Food Fraud: A Global Threat with Public Health and Economic Consequences*; Hellberg, R.S., Everstine, K., Sklare, S.A., Eds.; Academic Press: Cambridge, MA, USA, 2021; pp. 281–308.
104. Lozano-Castellón, J.; López-Yerena, A.; Domínguez-López, I.; Siscart-Serra, A.; Fraga, N.; Sámano, S.; López-Sabater, C. Extra Virgin Olive Oil: A Comprehensive Review of Efforts to Ensure Its Authenticity, Traceability, and Safety. *Compr. Rev. Food Sci. Food Saf.* **2022**, *21*, 2639–2664. [CrossRef] [PubMed]
105. Guillaume, C.; Miller, P.; Cantrill, R. Fraud in Fats and Oils. In *Food Fraud: A Global Threat with Public Health and Economic Consequences*; Hellberg, R.S., Everstine, K., Sklare, S.A., Eds.; Academic Press: Cambridge, MA, USA, 2021; pp. 151–175.
106. García, N.; Schwarzinger, S. Honey Fraud. In *Food Fraud: A Global Threat with Public Health and Economic Consequences*; Hellberg, R.S., Everstine, K., Sklare, S.A., Eds.; Academic Press: Cambridge, MA, USA, 2021; pp. 309–334.

107. Trifković, J.; Andrić, F.; Ristivojević, P.; Guzelmeric, E.; Yesilada, E. Analytical Methods in Tracing Honey Authenticity. *J. AOAC Int.* **2017**, *100*, 827–839. [CrossRef]
108. Álvarez-Suárez, J.M.; Majtan, J. Beyond fraud: Functional authenticity as the next frontier in honey quality assurance. *Food Control* **2026**, *181*, 111762. [CrossRef]
109. Lianou, A.; Papakonstantinou, M.; Nychas, G.-J.E.; Stoitsis, J. Fraud in meat and poultry products. In *Food Fraud: A Global Threat with Public Health and Economic Consequences*; Hellberg, R.S., Everstine, K., Sklare, S.A., Eds.; Academic Press: Cambridge, MA, USA, 2021; pp. 85–108.
110. Silva, A.J.; Hellberg, R.S.; Hanner, R.H. Seafood fraud. In *Food Fraud: A Global Threat with Public Health and Economic Consequences*; Hellberg, R.S., Everstine, K., Sklare, S.A., Eds.; Academic Press: Cambridge, MA, USA, 2021; pp. 109–137.
111. Gimonkar, S.; Van Fleet, E.E.; Boys, K.A. Dairy product fraud. In *Food Fraud: A Global Threat with Public Health and Economic Consequences*; Hellberg, R.S., Everstine, K., Sklare, S.A., Eds.; Academic Press: Cambridge, MA, USA, 2021; pp. 249–279. [CrossRef]
112. Ferreira, M.M.; Marins-Gonçalves, L.; De Souza, D. An integrative review of analytical techniques used in food authentication: A detailed description for milk and dairy products. *Food Chem.* **2024**, *457*, 140206. [CrossRef] [PubMed]
113. Abedini, A.; Salimi, M.; Mazaheri, Y.; Sadighara, P.; Alizadeh Sani, M.; Assadpour, E.; Jafari, S.M. Assessment of cheese frauds, and relevant detection methods: A systematic review. *Food Chem. X* **2023**, *19*, 100825. [CrossRef]
114. Samanidou, V.; Karageorgou, E.; Kourti, P. The Role of HPLC in the Authenticity Control of Milk and Dairy Products. In *Liquid Chromatography: Principles, Technology and Applications*; Ramos, F., Ed.; Nova Science Publishers: Hauppauge, NY, USA, 2013; pp. 173–206.
115. Joolaei Ahranjani, P.; Joolaei Ahranjani, P.; Dehghan, K.; Esfandiari, Z.; Ferrentino, G. Advances in Spectroscopic Techniques for the Detection of Cheese Adulteration: A Systematic Review. *Food Chem. X* **2025**, *29*, 102685. [CrossRef]
116. Perini, M.; Bontempo, L. Liquid chromatography coupled to isotope ratio mass spectrometry (LC-IRMS): A review. *Trends Anal. Chem.* **2022**, *147*, 116515. [CrossRef]
117. Liu, H.; Nie, J.; Liu, Y.; Wadood, S.A.; Rogers, K.M.; Yuan, Y.; Gan, R.-Y. A review of recent compound-specific isotope analysis studies applied to food authentication. *Food Chem.* **2023**, *415*, 135791. [CrossRef] [PubMed]
118. van Leeuwen, K.A.; Prenzler, P.D.; Ryan, D.; Camin, F. Gas Chromatography-Combustion-Isotope Ratio Mass Spectrometry for Traceability and Authenticity in foods and beverages. *Compr. Rev. Food Sci. Food Saf.* **2014**, *13*, 814–837. [CrossRef]
119. Park, J.; Cho, Y.S.; Seo, D.W.; Choi, J.Y. An update on the sample preparation and analytical methods for synthetic food colorants in food products. *Food Chem.* **2024**, *459*, 140333. [CrossRef]
120. Tzachristas, A.; Dasenaki, M.; Aalizadeh, R.; Thomaidis, N.S.; Proestos, C. LC-MS based metabolomics for the authentication of selected Greek white wines. *Microchem. J.* **2021**, *169*, 106543. [CrossRef]
121. Geana, E.I.; Popescu, R.; Costinel, D.; Dinca, O.R.; Stefanescu, I.; Ionete, R.E.; Bala, C. Verifying the red wines adulteration through isotopic and chromatographic investigations coupled with multivariate statistic interpretation of the data. *Food Control* **2016**, *62*, 1–9. [CrossRef]
122. Mac, H.X.; Pham, T.T.; Ha, N.T.T.; Nguyen, L.L.P.; Baranyai, L.; Friedrich, L. Current Techniques for Fruit Juice and Wine Adulterant Detection and Authentication. *Beverages* **2023**, *9*, 84. [CrossRef]
123. Yeganeh-Zare, S.; Farhadi, K.; Amiri, S. Rapid detection of apple juice concentrate adulteration with date concentrate, fructose and glucose syrup using HPLC-RID incorporated with chemometric tools. *Food Chem.* **2022**, *370*, 131015. [CrossRef]
124. Zhou, Q.; Li, G.; Ou-Yang, Z.; Yi, X.; Huang, L.; Wang, H. Volatile Organic Compounds Profiles to Determine Authenticity of Sweet Orange Juice Using Head Space Gas Chromatography Coupled with Multivariate Analysis. *Foods* **2020**, *9*, 505. [CrossRef]
125. Cuevas, F.J.; Moreno-Rojas, J.M.; Ruiz-Moreno, M.J. Assessing a Traceability Technique in Fresh Oranges (*Citrus sinensis* L. Osbeck) with an HS-SPME-GC-MS Method: Towards a Volatile Characterisation of Organic Oranges. *Food Chem.* **2017**, *221*, 1930–1938. [CrossRef]
126. Strojnik, L.; Stopar, M.; Zlatič, E.; Kokalj, D.; Naglič Gril, M.; Ženko, B.; Žnidaršič, M.; Bohanec, M.; Boshkovska, B.M.; Luštrek, M.; et al. Authentication of key aroma compounds in apple using stable isotope approach. *Food Chem.* **2019**, *277*, 766–773. [CrossRef]
127. Steingass, C.B.; Langen, J.; Carle, R.; Schmarr, H.-G. Authentication of Pineapple (*Ananas comosus* [L.] Merr.) Fruit Maturity Stages by Quantitative Analysis of γ- and δ-Lactones Using Headspace Solid-Phase Microextraction and Chirospecific Gas Chromatography-Selected Ion Monitoring Mass Spectrometry (HS-SPME-GC-SIM-MS). *Food Chem.* **2015**, *168*, 496–503. [CrossRef]
128. Ghaste, M.; Narduzzi, L.; Carlin, S.; Vrhovsek, U.; Shulaev, V.; Mattivi, F. Chemical Composition of Volatile Aroma Metabolites and Their Glycosylated Precursors That Can Uniquely Differentiate Individual Grape Cultivars. *Food Chem.* **2015**, *188*, 309–319. [CrossRef] [PubMed]
129. Zappi, A.; Marassi, V.; Kassouf, N.; Giordani, S.; Pasqualucci, G.; Garbini, D.; Roda, B.; Zattoni, A.; Reschiglian, P.; Melucci, D. A Green Analytical Method Combined with Chemometrics for Traceability of Tomato Sauce Based on Colloidal and Volatile Fingerprinting. *Molecules* **2022**, *27*, 5507. [CrossRef]
130. Kissoudi, M.; Sarakatsianos, I.; Samanidou, V. Isolation and purification of food-grade C-phycocyanin from *Arthrospira platensis* and its determination in confectionery by HPLC with diode array detection. *J. Sep. Sci.* **2018**, *41*, 975–981. [CrossRef] [PubMed]

https://doi.org/10.3390/separations13010039

131. Papapetros, S.; Louppis, A.; Kosma, I.; Kontakos, S.; Badeka, A. Characterization and Differentiation of Botanical and Geographical Origin of Selected Popular Sweet Cherry Cultivars Grown in Greece. *J. Food Compos. Anal.* **2018**, *72*, 48–56. [CrossRef]
132. Dinis, K.; Tsamba, L.; Jamin, E.; Camel, V. Untargeted Metabolomics-Based Approach Using UHPLC-HRMS to Authenticate Carrots (*Daucus carota* L.) Based on Geographical Origin and Production Mode. *Food Chem.* **2023**, *423*, 136273. [CrossRef]
133. Lim, D.K.; Long, N.P.; Mo, C.; Dong, Z.; Cui, L.; Kim, G.; Kwon, S.W. Combination of mass spectrometry-based targeted lipidomics and supervised machine learning algorithms in detecting adulterated admixtures of white rice. *Food Res. Int.* **2017**, *100*, 814–821. [CrossRef]
134. Xiao, R.; Ma, Y.; Zhang, D.; Qian, L. Discrimination of conventional and organic rice using untargeted LC-MS-based metabolomics. *J. Cereal Sci.* **2018**, *82*, 73–81. [CrossRef]
135. Bönick, J.; Huschek, G.; Rawel, H.M. Determination of wheat, rye and spelt authenticity in bread by targeted peptide biomarkers. *J. Food Compos. Anal.* **2017**, *58*, 82–91. [CrossRef]
136. Weesepoel, Y.; Heenan, S.; Boerrigter-Eenling, R.; Venderink, T.; Blokland, M.; van Ruth, S. Protocatechuic Acid Levels Discriminate Between Organic and Conventional Wheat from Denmark. *Chimia* **2016**, *70*, 360–363. [CrossRef]
137. Du, Q.W.; Zhu, M.T.; Shi, T.; Luo, X.; Gan, B.; Tang, L.J.; Chen, Y. Adulteration detection of corn oil, rapeseed oil and sunflower oil in camellia oil by in situ diffuse reflectance near-infrared spectroscopy and chemometrics. *Food Control* **2021**, *121*, 107577. [CrossRef]
138. Zhou, X.; Zhang, Q.; Chen, X.; Li, X.; Han, C. In-situ assessment of olive oil adulteration with soybean oil based on thermogravimetric-gas chromatography/mass spectrometry combined with chemometrics. *Food Control* **2021**, *130*, 108251. [CrossRef]
139. Tian, H.; Wu, D.; Chen, B.; Yuan, H.; Yu, H.; Lou, X.; Chen, C. Rapid identification and quantification of vegetable oil adulteration in raw milk using a flash gas chromatography electronic nose combined with machine learning. *Food Control* **2023**, *150*, 109758. [CrossRef]
140. Ratnasekhar, C.H.; Khan, S.; Rai, A.K.; Mishra, H.; Verma, A.K.; Lal, R.K.; Ananda Kumar, T.M.; Elliott, C.T. Rapid metabolic fingerprinting meets machine learning models to identify authenticity and detect adulteration of essential oils with vegetable oils: Mentha and Ocimum study. *Food Chem.* **2025**, *471*, 142709. [CrossRef]
141. Zou, Z.; Wang, Q.; Li, M.; Zhen, J.; Yuan, D.; Zhang, Q.; Cui, Q.; Li, M.; Zhou, M.; Yin, S.; et al. Combination of gas chromatography-mass spectrometry and hyperspectral imaging for identification of adulterated Safflower seed oil. *J. Food Compos. Anal.* **2024**, *135*, 106593. [CrossRef]
142. Aghili, N.S.; Rasekh, M.; Karami, H.; Azizi, V.; Gancarz, M. Detection of fraud in sesame oil with the help of artificial intelligence combined with chemometrics methods and chemical compounds characterization by gas chromatography–mass spectrometry. *LWT* **2022**, *167*, 113863. [CrossRef]
143. Kozub-Pędrak, A.; Nikolaichuk, H.; Przykaza, K.; Sumara, A.; Pitucha, M.; Tomaszewska-Gras, J.; Fornal, E. LC-MS-based glyceride profiling of cold-pressed fruit seed oils for effective authenticity testing. *Food Chem.* **2025**, *493*, 146039. [CrossRef]
144. Egido, C.; Saurina, J.; Sentellas, S.; Núñez, O. Honey fraud detection based on sugar syrup adulterations by HPLC-UV fingerprinting and chemometrics. *Food Chem.* **2024**, *436*, 137758. [CrossRef]
145. Jakanovski, M.V.; Horvacki, N.M.; Dramićanin, A.M.; Radoičić, A.D.; Šegan, S.B.; Milojković-Opsenica, D.M. Organic acids as promising authenticity markers of *Robinia pseudoacacia* honey. *J. Food Compos. Anal.* **2025**, *148*, 108384. [CrossRef]
146. Mostoles, D.; Egido, C.; Mara, A.; Sanna, G.; Sentellas, S.; Saurina, J.; Núñez, O. Potential of high-performance liquid chromatography with ultraviolet detection (HPLC-UV) fingerprints to assess the geographical production origin and authenticity of honey. *Microchem. J.* **2025**, *209*, 112669. [CrossRef]
147. Duft, R.G.; Griffin, J.L.; Stead, D.A. MEATiCode: A comprehensive proteomic LC-MS/MS method for simultaneous species identification in meat authentication. *Food Chem.* **2025**, *483*, 144231. [CrossRef]
148. Kang, C.; Zhang, Y.; Zhang, M.; Qi, J.; Zhao, W.; Gu, J.; Guo, W.; Li, Y. Screening of specific quantitative peptides of beef by LC-MS/MS coupled with OPLS-DA. *Food Chem.* **2022**, *387*, 132932. [CrossRef]
149. Łuciuk, A.; Paszkowska, E.; Sumara, A.; Kozub-Pędrak, A.; Wielgosz, A.; Stachniuk, A.; Montowska, M.; Fornal, E. New approaches towards fish authentication: Pollock and hake differentiation by liquid chromatography high resolution mass spectrometry. *Food Control* **2025**, *171*, 111085. [CrossRef]
150. Wang, X.; Liu, Y.; Zhu, S.; Bai, Y.; Li, C.; Guo, J. Authenticity of domestic animal milk: An identification strategy by supercritical fluid chromatography quadruple time-of-flight mass spectrometry. *Food Chem.* **2025**, *480*, 143940. [CrossRef]
151. Nguyen, A.V.; Vu, A.T.N.; Deineka, V.I.; Deineka, L.A.; Da, T.K.T.L. Novel approach for determination of milk fat adulteration with non-milk fat by RP-HPLC. *J. Food Compos. Anal.* **2023**, *121*, 105419. [CrossRef]
152. Barbu, I.M.; Lamers, R.-J.A.N.; Gerritsen, H.W.; Blokland, M.H.; Bremer, M.G.E.G.; Alewijn, M. Endogenous protein and peptide analysis with LC-MS/(MS): A feasibility study for authentication of raw-milk farmer's cheese. *Int. Dairy J.* **2021**, *117*, 104990. [CrossRef]

https://doi.org/10.3390/separations13010039

153. Kritikou, A.S.; Aalizadeh, R.; Damalas, D.E.; Barla, I.V.; Baessmann, C.; Thomaidis, N.S. MALDI-TOF-MS integrated workflow for food authenticity investigations: An untargeted protein-based approach for rapid detection of PDO feta cheese adulteration. *Food Chem.* **2022**, *370*, 131057. [CrossRef] [PubMed]
154. Sun, H.; Lu, W.; Gao, B. Non-targeted detection of butter adulteration using pointwise UHPLC-ELSD and UHPLC-UV fingerprints with chemometrics. *Food Chem.* **2021**, *356*, 129604. [CrossRef]
155. Silva, L.K.R.; de Jesus, J.C.; Onelli, R.R.V.; Conceição, D.G.; Santos, L.S.; Ferrão, S.P.B. Spectroscopy (MIR), chromatography (RP-HPLC) and chemometrics applied to soluble peptides to discriminate the geographic origin of coalho cheese. *Biocatal. Agric. Biotechnol.* **2023**, *50*, 102678. [CrossRef]
156. European Parliament and Council. *Regulation (EC) No 852/2004 on the Hygiene of Foodstuffs*; European Parliament and Council: Brussels, Belgium, 2004.
157. European Parliament and Council. *Regulation (EC) No 1935/2004 on Materials and Articles Intended to Come into Contact with Food and Repealing Directives 80/590/EEC and 89/109/EEC*; European Parliament and Council: Brussels, Belgium, 2004.
158. European Commission. *Commission Regulation (EU) No 10/2011 on Plastic Materials and Articles Intended to Come into Contact with Food*; European Parliament and Council: Brussels, Belgium, 2011.
159. GFSI. *Chemicals in Food Hygiene—Volume 2: Cleaning Agents, Sanitisers and Disinfectants in Food Businesses: Detection of Traces and Human Risk Assessment Processes*; GFSI—The Consumer Goods Forum: Levallois-Perret, France, 2019.
160. Wrona, M.; Nerín, C. Analytical Approaches for Analysis of Safety of Modern Food Packaging: A Review. *Molecules* **2020**, *25*, 752. [CrossRef] [PubMed]
161. Nerín, C.; Canellas, E.; Vera, P. Migration of packaging and labeling components and advances in analytical methodology supporting exposure assessment. In *Present Knowledge in Food Safety, A Risk-Based Approach Through the Food Chain*; Knowles, M.E., Anelich, L.E., Boobis, A.R., Popping, B., Eds.; Academic Press: London, UK, 2023; pp. 218–239. [CrossRef]
162. Lestido-Cardama, A.; Barbosa-Pereira, L.; Sendón, R.; Bustos, J.; Paseiro Losada, P.; Rodríguez Bernaldo de Quirós, A. Chemical safety and risk assessment of bio-based and/or biodegradable polymers for food contact: A review. *Food Res. Int.* **2025**, *202*, 115737. [CrossRef]
163. Spinei, M.; Oroian, M. Identification and quantification of intentionally and non-intentionally added substances occurring from biodegradable food packaging: A review. *Food Res. Int.* **2025**, *218*, 116880. [CrossRef]
164. Rodríguez-Ramos, R.; Santana-Mayor, A.; Herrera-Herrera, A.V.; Socas-Rodríguez, B.; Rodríguez-Delgado, M.A. Recent advances in the analysis of plastic migrants in food. *TrAC Trends Anal. Chem.* **2024**, *178*, 117847. [CrossRef]
165. Carrero-Carralero, C.; Escobar-Arnanz, J.; Ros, M.; Jiménez-Falcao, S.; Sanz, M.L.; Ramos, L. An untargeted evaluation of the volatile and semi-volatile compounds migrating into food simulants from polypropylene food containers by comprehensive two-dimensional gas chromatography-time-of-flight mass spectrometry. *Talanta* **2019**, *195*, 800–806. [CrossRef]
166. Tang, C.; Gómez Ramos, M.J.; Heffernan, A.; Kaserzon, S.; Rauert, C.; Lin, C.-Y.; Mueller, J.F.; Wang, X. Evaluation and identification of chemical migrants leached from baby food pouch packaging. *Chemosphere* **2023**, *340*, 139758. [CrossRef] [PubMed]
167. Bonini, M.; Errani, E.; Zerbinati, G.; Ferri, E.; Girotti, S. Extraction and gas chromatographic evaluation of plasticizers content in food packaging films. *Microchem. J.* **2008**, *90*, 31–36. [CrossRef]
168. Kavalchuk, Y.; Shishov, A.; Pochivalov, A.; Safonova, E.; Bulatov, A. Primary alcohol-induced coacervation in alkyl polyglucoside micellar solution for supramolecular solvent-based microextraction and chromatographic determination of phthalates in baby food. *Talanta* **2024**, *280*, 126748. [CrossRef] [PubMed]
169. Horodytska, O.; Cabanes, A.; Fullana, A. Non-intentionally added substances (NIAS) in recycled plastics. *Chemosphere* **2020**, *251*, 126373. [CrossRef]
170. Lin, J.; Wu, W.-L.; Zhong, A.; Xian, Y.-P.; Zhong, H.-N.; Dong, B.; Liang, M.; Hu, J.-P.; Wu, Y.-N.; Yang, X.-F.; et al. Non-targeted analysis and risk assessment of intentionally and non-intentionally added substances migrating from the emerging biodegradable food contact material poly(butylene adipate-co-terephthalate)/modified starch blend film. *Food Packag. Shelf Life* **2023**, *40*, 101190. [CrossRef]
171. Zhong, H.; Li, Z.; Chen, S.; Zeng, Y.; Zheng, J.; Zeng, Y.; Li, D. Simultaneous Quantitative Analysis of Six Isothiazolinones in Water-Based Adhesive Used for Food Contact Materials by High-Performance Liquid Chromatography–Tandem Mass Spectrometry (HPLC–MS/MS). *Molecules* **2019**, *24*, 3894. [CrossRef]
172. Guan, W.; He, Y.; McClements, D.J.; Chen, J.; Ma, D. Risk assessment of migrants released from multilayer packaging materials: Direct immersion-solid-phase microextraction coupled to gas chromatography-mass spectrometry. *Food Packag. Shelf Life* **2024**, *46*, 101407. [CrossRef]
173. Sapozhnikova, Y. Non-targeted screening of chemicals migrating from paper-based food packaging by GC-Orbitrap mass spectrometry. *Talanta* **2021**, *226*, 122120. [CrossRef]

https://doi.org/10.3390/separations13010039

174. Blanco-Zubiaguirre, L.; Zabaleta, I.; Usobiaga, A.; Prieto, A.; Olivares, M.; Zuloaga, O.; Elizalde, M.P. Target and suspect screening of substances liable to migrate from food contact paper and cardboard materials using liquid chromatography-high resolution tandem mass spectrometry. *Talanta* **2020**, *208*, 120394. [CrossRef]
175. Moret, S.; Conchione, C.; Vanini, R.; Barp, L. Mineral oil contamination along the cocoa supply chain: Impact of storage and transportation in jute bags and assessment of contamination transfer into cocoa powder and cocoa butter. *Food Chem.* **2025**, *496*, 146955. [CrossRef]
176. Biedermann, M.; Munoz, C.; Grob, K. Update of on-line coupled liquid chromatography—gas chromatography for the analysis of mineral oil hydrocarbons in foods and cosmetics. *J. Chromatogr. A* **2017**, *1521*, 140–149. [CrossRef]
177. Groschke, M.; Becker, R. Comparison of carrier gases for the separation and quantification of mineral oil hydrocarbon (MOH) fractions using online coupled high performance liquid chromatography-gas chromatography-flame ionisation detection. *J. Chromatogr. A* **2024**, *1726*, 464946. [CrossRef]
178. Van Asselt, A.J.; Huellemeier, H.A.; Schuten, H.J.; Augustin, W. Is cleaning in place (CIP) ready for the future? —A critical assessment. *Food Bioprod. Process.* **2025**, *154*, 270–281. [CrossRef]
179. Akhare, H.; Rajput, N.; Jadav, T.; Sengupta, P. Development of a comprehensive cleaning and detoxification approach for selected sodium-glucose cotransporter-2 inhibitors including establishment of highly sensitive analytical method for their simultaneous surface residue estimation. *Microchem. J.* **2025**, *209*, 112587. [CrossRef]
180. Twomey, L.; Furey, A.; O'Brien, T.; Beresford, T.; O'Brien, B.; Gleeson, D. An evaluation of dairy product quality where chlorine-free cleaning is employed across the dairy processing chain from bacterial and chemical residue perspectives. *Int. Dairy J.* **2023**, *146*, 105739. [CrossRef]
181. Kürzl, C.; Wohlschläger, H.; Schiffer, S.; Kulozik, U. Concentration, purification and quantification of milk protein residues following cleaning processes using a combination of SPE and RP-HPLC. *MethodsX* **2022**, *9*, 101695. [CrossRef]
182. Li, J.; Yan, J.; Cao, Q.; Wu, D.; Lin, W.; Jin, M. Determination of residual 15 quaternary ammonium compounds in dairy products by dispersive solid-phase extraction purification with ultra-high performance liquid chromatography-tandem mass spectrometry. *J. Chromatogr. A* **2026**, *1765*, 466525. [CrossRef]
183. Slimani, K.; Pirotais, Y.; Maris, P.; Abjean, J.-P.; Hurtaud-Pessel, D. Liquid chromatography–tandem mass spectrometry method for the analysis of N-(3-aminopropyl)-N-dodecylpropane-1,3-diamine, a biocidal disinfectant, in dairy products. *Food Chem.* **2018**, *262*, 168–177. [CrossRef] [PubMed]
184. Slimani, K.; Féret, A.; Pirotais, Y.; Maris, P.; Abjean, J.-P.; Hurtaud-Pessel, D. Liquid chromatography-tandem mass spectrometry multiresidue method for the analysis of quaternary ammonium compounds in cheese and milk products: Development and validation using the total error approach. *J. Chromatogr. A* **2017**, *1517*, 86–96. [CrossRef] [PubMed]
185. dos Santos, I.D.; Zomer, P.; Pizzutti, I.R.; Wagner, R.; Mol, H. Multi-residue determination of biocides in dairy products and slurry feed using QuEChERS extraction and liquid chromatography combined with high resolution mass spectrometry (LC-ESI-QOrbitrap™-MS). *Food Chem.* **2024**, *457*, 140117. [CrossRef] [PubMed]
186. Cortés-Bautista, S.; Navarro-Utiel, R.; Ballester-Caudet, A.; Campíns-Falcó, P. Towards in field miniaturized liquid chromatography: Biocides in wastewater as a proof of concept. *J. Chromatogr. A* **2022**, *1673*, 463119. [CrossRef]
187. Hakakzade, E.; Mollahosseini, A. Application of reusable flat-membrane in electro-membrane extraction for tamsulosin hydrochloride determination in cleaning validation samples of sterile production line equipment by RP-HPLC. *Eur. J. Pharm. Sci.* **2021**, *161*, 105793. [CrossRef] [PubMed]
188. Prieto-Blanco, M.C.; Moliner-Martinez, Y.; Campíns-Falcó, P. Combining poly(dimethyldiphenylsiloxane) and nitrile phases for improving the separation and quantitation of benzalkonium chloride homologues: In-tube solid phase microextraction–capillary liquid chromatography–diode array detection-mass spectrometry for analyzing industrial samples. *J. Chromatogr. A* **2013**, *1297*, 226–230. [CrossRef] [PubMed]
189. Zhang, M.; Lian, K.; Ai, L.; Kang, W.; Zhao, T. Simultaneous determination of 11 antiseptic ingredients in surface water based on polypyrrole decorated magnetic nanoparticles. *RSC Adv.* **2020**, *10*, 37473–37481. [CrossRef]
190. Haidar Ahmad, I.A.; Tam, J.; Li, X.; Duffield, W.; Tarara, T.; Blasko, A. Cleaning verification: Exploring the effect of the cleanliness of stainless steel surface on sample recovery. *J. Pharm. Biomed. Anal.* **2017**, *134*, 108–115. [CrossRef]
191. Lawate, S. Lubricants and Fluids for the Food Industry. In *Lubricant Additives*, 3rd ed.; CRC Press: Boca Raton, FL, USA, 2017; pp. 1–18.
192. Gokarn, N.A.; Kiran, K.N. Role of Lubricants in Industry: A Review. *J. ISAS* **2023**, *2*, 69–83. [CrossRef]
193. Moens, E. EHEDG Update: Production and Use of Food-Grade Lubricants. *Trends Food Sci. Technol.* **2003**, *14*, 157–162. [CrossRef]
194. Schaefer, A.; Küchler, T.; Simat, T.J.; Steinhart, H. Migration of lubricants from food packagings: Screening for lipid classes and quantitative estimation using normal-phase liquid chromatographic separation with evaporative light scattering detection. *J. Chromatogr. A* **2003**, *1017*, 107–116. [CrossRef]
195. Rudnick, L.R.; Erhan, S.Z. Natural Oils as Lubricants. In *Synthetics, Mineral Oils, and Bio-Based Lubricants*, 1st ed.; CRC Press: Boca Raton, FL, USA, 2005; pp. 1–8.

https://doi.org/10.3390/separations13010039

196. Salih, N.; Salimon, J. A Review on New Trends, Challenges and Prospects of Ecofriendly Green Food-Grade Biolubricants. *Biointerface Res. Appl. Chem.* **2022**, *12*, 1185–1207. [CrossRef]
197. Fiselier, K.; Grob, K. Determination of Mineral Oil Paraffins in Foods by On-Line HPLC–GC–FID: Lowered Detection Limit; Contamination of Sunflower Seeds and Oils. *Eur. Food Res. Technol.* **2009**, *229*, 679–688. [CrossRef]
198. Li, B.; Wu, Y.; Liu, L.; Ouyang, J.; Ren, J.; Wang, Y.; Wang, X. Determination of Mineral Oil-Saturated Hydrocarbons (MOSH) in Vegetable Oils by Large Scale Off-Line SPE Combined with GC-FID. *J. Am. Oil Chem. Soc.* **2017**, *94*, 215–223. [CrossRef]
199. Shang, D.; Park, G.; Brunswick, P.; van Aggelen, G.; Yang, C.; Kim, M. Combined Gas and Liquid Chromatography Tandem Mass Spectrometry Applications for Forensic Lubricant and Vegetable Oil Spill Identification. In *Oil Spill Environmental Forensics Case Studies*; Stout, S.A., Wang, Z., Eds.; Butterworth-Heinemann: Oxford, UK, 2018; pp. 117–130. [CrossRef]
200. Kohler, M.; Heeb, N.V. Characterization of ageing products of ester-based synthetic lubricants by liquid chromatography with electrospray ionization mass spectrometry and by electrospray ionization (tandem) mass spectrometry. *J. Chromatogr. A* **2001**, *926*, 161–165. [CrossRef]
201. Joint Research Centre (European Commission). *Guidance on Sampling, Analysis and Data Reporting for the Monitoring of Mineral Oil Hydrocarbons in Food and Food Contact Materials, in the Frame of Commission Recommendation (EU) 2017/84*; European Commission: Luxembourg, 2023; Available online: https://publications.jrc.ec.europa.eu/repository/handle/JRC133174 (accessed on 1 December 2025).
202. Kreisberger, G.; Himmelsbach, M.; Buchberger, W.; Klampfl, C.W. Identification and semi-quantitative determination of antioxidants in lubricants employing thin-layer chromatography-spray mass spectrometry. *J. Chromatogr. A* **2015**, *1383*, 169–174. [CrossRef] [PubMed]
203. Qi, Y.; Zhao, P. Synthetic Pyrethroid residues in food: Chemical behavior, analytical advances, and safety perspectives. *Food Chem.* **2025**, *490*, 145100. [CrossRef]
204. Wang, Y.; Fu, Y.; Wang, Y.; Lu, Q.; Ruan, H.; Luo, J.; Yang, M. A comprehensive review on the pretreatment and detection methods of neonicotinoid insecticides in food and environmental samples. *Food Chem. X* **2022**, *15*, 100375. [CrossRef]
205. Wang, Q.; Wang, F.; Huang, Z.; Zheng, J.; Liang, X.; Han, Y.; Ju, X.; Liu, X.; Zhang, H. Detection methods of diamide pesticides in food: Challenges and future directions. *Food Chem.* **2025**, *493*, 145760. [CrossRef]
206. Aghris, S.; Tahiri Alaoui, O.; Laghrib, F.; Farahi, A.; Bakasse, M.; Saqrane, S.; Lahrich, S.; El Mhammedi, M.A. Extraction and determination of flubendiamide insecticide in food samples: A review. *Curr. Res. Food Sci.* **2022**, *5*, 401–413. [CrossRef]
207. Liu, Y.; Ma, C.; Chen, M.; Hao, J.; Liu, W.; Yu, J. Hydrophobic deep eutectic solvent-based ferrofluid for the multi-residue analysis of pyrethroid insecticides in agricultural products: Facile pretreatment and MS-friendly detection. *J. Chromatogr. A* **2025**, *1759*, 466262. [CrossRef] [PubMed]
208. Hai, X.; Niu, J.; Ren, T.; Fu, R.; Li, H.; Zhang, Y.; Guo, X.; Di, X. Rhamnolipids-based bio-supramolecular solvents as green and sustainable media for extraction of pyrethroid insecticides in water and food matrices. *J. Chromatogr. A* **2024**, *1731*, 465215. [CrossRef]
209. Jaganmohanrao, L. Role of deep eutectic solvents as alternate solvents in the microextraction and estimation of pesticide, insecticide, fungicide residues and metal contaminants in tea (*Camellia sinensis*). *Microchem. J.* **2025**, *208*, 112515. [CrossRef]
210. Gallocchio, F.; Basilicata, L.; Benetti, C.; Angeletti, R.; Binato, G. Multi-residue determination of eleven anticoagulant rodenticides by high-performance liquid chromatography with diode array/fluorimetric detection: Investigation of suspected animal poisoning in the period 2012–2013 in north-eastern Italy. *Forensic Sci. Int.* **2014**, *244*, 63–69. [CrossRef]
211. Okoniewski, R.; Neely, S.; Denn, M.; Djatsa, A.; Tran, B.N. Rapid method for the detection of rodenticides in contaminated foods. *J. Chromatogr. B* **2021**, *1186*, 123005. [CrossRef]
212. Fourel, I. Enantiomer fraction evaluation of the four stereoisomers of second-generation anticoagulant rodenticides in biological matrices with polysaccharide-based chiral selectors and liquid chromatography tandem mass spectrometry. *J. Chromatogr. A* **2022**, *1676*, 463209. [CrossRef]
213. Berset, J.D.; Mermer, S.; Robel, A.E.; Walton, V.M.; Chien, M.L.; Field, J.A. Direct residue analysis of systemic insecticides and some of their relevant metabolites in wines by liquid chromatography–mass spectrometry. *J. Chromatogr. A* **2017**, *1506*, 45–54. [CrossRef] [PubMed]
214. Bhamdare, H.; Pahade, P.; Bose, D.; Durgbanshi, A.; Carda-Broch, S.; Peris-Vicente, J. Evaluating the effectiveness of different household washing techniques for removal of insecticides from spinach and chickpea leaves by micellar liquid chromatography. *J. Chromatogr. A* **2024**, *1730*, 465043. [CrossRef] [PubMed]
215. Parven, A.; Khan, M.S.I.; Prodhan, M.D.H.; Venkateswarlu, K.; Megharaj, M.; Meftaul, I.M. Human health risk assessment through quantitative screening of insecticide residues in two green beans to ensure food safety. *J. Food Compos. Anal.* **2021**, *103*, 104121. [CrossRef]
216. Chang, H.-Y.; Yang, W.-C.; Xue, Y.-J.; Tsai, M.-Y.; Wang, J.-H.; Chang, G.-R. Phthalates and organophosphorus insecticide residues in shrimp determined by liquid/gas chromatography–tandem mass spectrometry and a health risk assessment. *Mar. Pollut. Bull.* **2019**, *144*, 140–145. [CrossRef] [PubMed]

https://doi.org/10.3390/separations13010039

217. Lofty, H.M.; Abd El-Aziz, A.A.; Monir, H.H. Determination of insecticides malathion and lambda-cyhalothrin residues in zucchini by gas chromatography. *Bull. Fac. Pharm. Cairo Univ.* **2013**, *51*, 255–260. [CrossRef]
218. Bai, B.; Guo, Y.; Meng, S.; Gong, Y.; Bo, T.; Zhang, J.; Shen, D.; Fan, S.; Yang, Y. Determination of insecticide residues in beverages based on MIL-100(Fe) dispersive solid-phase microextraction in combination with dispersive liquid-liquid microextraction followed by HPLC-MS/MS. *Food Chem.* **2024**, *453*, 139660. [CrossRef]
219. Pan, S.; Qiu, Q.; Li, X.; Huang, Y.; Shang, C.; Cao, X. An automated magnetic solid-phase extraction method for simultaneous determination of 22 nicotinic insecticides and fungicides in peach samples combined with ultra-high performance liquid chromatography-high resolution mass spectrometry. *J. Food Compos. Anal.* **2025**, *139*, 107160. [CrossRef]
220. Sivasankari, S.; Suganthi, A.; Elaiyabharathi, T.; Shanmugam, P.S.; Renukadevi, P.; Veeranan, G.V.A.; Venkidusamy, M.; Palanisamy, K.; Pandi, T.; Venkatachalam, S. Study on impact of decontamination and processing methods on insecticide residues in green and red chilli (*Capsicum annum* L.) using LC-MS/MS. *Food Control* **2025**, *169*, 110999. [CrossRef]
221. FSSC. *FSSC 22000 Scheme Food Safety Management Certification Version 6*; FSSC: Wageningen, The Netherlands, 2023.
222. BRCGS. *BRCGS Global Standard for Food Safety, Issue 9*; BRCGS: London, UK, 2022.
223. IFS Management GmbH. *IFS Food Version 8: Standard for Auditing Product and Process Compliance in Relation to Food Safety and Quality*; IFS Management GmbH: Berlin, Germany, 2023.

https://doi.org/10.3390/separations13010039

separations

MDPI

Article

Capsule Phase Microextraction Combined with Chemometrics for the HPLC Determination of Amphotericin B in Human Serum

Natalia Manousi [1], Anastasia Korpeti [1], Abuzar Kabir [2], Kenneth G. Furton [2] and Constantinos K. Zacharis [1,*]

[1] Laboratory of Pharmaceutical Analysis, Department of Pharmaceutical Technology, School of Pharmacy, Aristotle University of Thessaloniki, GR-54124 Thessaloniki, Greece
[2] Department of Chemistry and Biochemistry, Florida International University, Miami, FL 33199, USA
* Correspondence: czacharis@pharm.auth.gr; Tel.: +30-23-1099-7663

Abstract: This article discusses the use of a sorbent-based microextraction technique employing a capsule device to isolate amphotericin B (AMB) from human serum before analysis by high performance liquid chromatography (HPLC). AMB is a macrocyclic compound used for the treatment of invasive fungal infections. Before determining AMB in human serum by HPLC, a sample preparation step is required. Capsule phase microextraction (CPME) integrates the stirring and filtration mechanisms in a single unit, simplifying the sample preparation procedure. Moreover, it results in fast extraction kinetics and high extraction efficiency, while it has proved to be a powerful tool for bioanalysis. Different sol–gel sorbent encapsulated microextraction capsules were investigated, and sol–gel Carbowax 20 M was finally chosen as the basis for the microextraction device. Accordingly, the sample preparation protocol was investigated using a face-centered central composite design to achieve good extraction performance. The optimum protocol was validated in terms of linearity, selectivity, limit of detection (LOD), limit of quantitation (LOQ), precision, and accuracy. The linear range of the developed approach was 0.10–10.0 μg mL^{-1}. The LOD value was 0.03 μg mL^{-1}, and the LOQ value was 0.10 μg mL^{-1}. Method accuracy (expressed as relative recovery) was 87–113%, while the relative standard deviation of the repeatability (s_r) and within-laboratory reproducibility (s_R) were <12.4%. The sol–gel sorbent encapsulated microextraction capsules were reusable for at least 10 extraction cycles. All things considered, the proposed method exhibited good overall performance, and it could be used in bioanalysis for quality control, therapeutic drug monitoring and research purposes.

Keywords: capsule phase microextraction; amphotericin B; HPLC; central composite design; validation; serum

1. Introduction

Amphotericin B (AMB) is a macrocyclic compound that contains both amino and carboxyl groups, and it shows significant biological activity due to its characteristic structure. In particular, AMB has an amphipathic character because of the hydroxyl groups on one side of the macrolide ring and the rigid lipophilic chain with seven conjugated double bonds on the other side [1]. This drug is widely used to treat invasive fungal infections, including coccidiosis, aspergillosis, candidiasis, and cryptococcosis [1,2]. The monitoring of the dosages and the distribution in the human body of AMB during therapy is an important parameter that can play a crucial role in the effectiveness of the treatment. Specifically, the careful monitoring of patients that undergo treatment with AMB is required since the long-term use of this drug is associated with various adverse effects, including infusion-related toxicity and nephrotoxicity [3]. Thus, the development of simple and efficient bioanalytical protocols for the monitoring of AMB is of utmost importance.

High-performance liquid chromatography is typically used for determining AMB in biological samples [1,3]. Sample preparation is considered a critical step of the analytical procedure that aims to remove potentially interfering compounds, preconcentrate the target analyte, and reduce the matrix effect [4]. The traditional sample preparation techniques used in bioanalysis include liquid–liquid extraction, solid-phase extraction, and protein precipitation [5,6]. However, these protocols are characterized by some significant disadvantages, including high consumption of samples and hazardous solvents, which contradict the principles of Green Analytical Chemistry (GAC) [7]. Depending on the required extent of analyte extraction, the required amount of sample and solvent and the extraction device's geometry, numerous microextraction methods have been proposed and used in bioanalysis. Typical examples of them include microextraction by packed sorbent [8], disposable pipette extraction [9], solid-phase microextraction [8], liquid-phase microextraction [10], stir-bar sorptive extraction [11], fabric phase sorptive extraction [12] and capsule phase microextraction (CPME) [13].

CPME was proposed by Kabir and Furton in 2017 [14]. In CPME, extraction of the target analytes from the sample matrix is performed by using specially developed microextraction capsules. These capsules are prepared by combining two porous polypropylene capillary tubes. The first tube contains the sol–gel sorbent, and the second includes a magnetic rod. As such, the stirring and extraction mechanisms are integrated into a versatile "all-in-one" device that is handy and reusable [15]. Moreover, the filtration mechanism is also integrated in the final device because of the inherent porosity of the tubes. The integration of sample preparation steps complies with the recently proposed 10 principles of Green Sample Preparation [16].

The application of CPME in bioanalysis results in rapid and miniaturized sample preparation schemes that are characterized by good performance. This can be attributed to the high efficiency of the sol–gel sorbents that are typically used. These sorbents are either coated in cotton fibers and inserted into the polypropylene tube or prepared in situ within the lumen of the tube [17]. Until now, CPME has been used in different bioanalytical protocols for the extraction of ibuprofen from urine [13], antifungal and anthelmintic drugs from urine [18], doxorubicin and its metabolites from rat plasma [17], and the extraction of bisphenols from human breast milk [15]. In all cases, the sample preparation scheme was characterized by handling simplicity, high sample throughput, reduced consumption of organic solvents, and reduced waste generation.

In this work, CPME was used for the first time for the determination of AMB in human serum samples. The final determination of the drug was conducted by high-performance liquid chromatography–ultraviolet detection (HPLC-UV). All factors that could potentially influence the performance of the bioanalytical protocol were studied, and chemometrics were employed for the optimization of the most critical ones. Following its optimization, the entire methodology was validated and used for the monitoring of AMB in human serum samples.

2. Materials and Methods

2.1. Reagents and Solutions

AMB (>98%), nimesulide (≥98.0%), methyl trimethoxysilane, tetramethyl orthosilicate, and poly(tetrahydrofuran) were obtained from Sigma-Aldrich (St. Louis, MO, USA). HPLC-grade solvents, such as methanol (MeOH) and acetonitrile (ACN), were obtained from Honeywell (Charlotte, NC, USA). Milli-Q water was provided by a B30 purification system (Adrona SIA, Riga, Latvia).

Stock standard solutions of AMB at a concentration of 1000 μg mL^{-1} were prepared in DMSO. Nimesulide (used as internal standard, ISTD) solution was prepared in MeOH by dissolving 25 mg of drug API in a 25 mL volumetric flask. The above solutions were kept at 4 °C. Working standards were made in water from the stock solutions.

Cylindrical magnetic rods (1/4″ × 1/16″) were obtained from K&J Magnetics Inc. (Pipersville, PA, USA). Membrane Accurel® porous membranes were obtained from 3M

Inc. (St. Paul, MN, USA). Isopropanol, CH_2Cl_2, NH_4OH and HCl were obtained from Fisher Scientific (Milwaukee, WI, USA).

The preparation of the microextraction capsules is described in detail elsewhere [13]. In brief, the general manufacturing procedure involved the following processes:

(a) Preparation of the porous tubular membranes.
(b) Preparation of the sol solution for the herein examined sol–gel sorbents.
(c) In situ creation of a monolithic bed within the membrane.
(d) Aging, conditioning, and cleaning of the obtained CPME media.

2.2. HPLC Instrumentation and Conditions

A Shimadzu 2010A HPLC-UV system (Kyoto, Japan) equipped with a high-pressure quaternary gradient pump, a column compartment, an UV detector, and an autosampler was utilized. LC Solutions software (vs. 1.25 SP4) was used. Separation of AMB and the ISTD was achieved using a Poroshell 120 EC-C18 column (50 × 4.6 mm, 2.7 μm) from Agilent Technologies (Santa Clara, CA, USA). The target analyte was separated from the ISTD and the matrix interferences in gradient mode using citric acid 22 mM (pH 4.2 adjusted with 1 M NaOH solution): MeOH 60:40 v/v (A) and citric acid 22 mM (pH 4.2 adjusted with 1 M NaOH solution): MeOH 20:80 v/v (B). The initial mobile phase composition was 50:50, A/B v/v. The composition was changed to 0:100 A/B v/v at 3 min, and it was kept constant until 5 min. Finally, it changed back to the initial parameters (i.e., 50:50, A/B v/v) at 5.5 min, and the system was equilibrated until 10 min. A flow rate of 0.6 mL min^{-1} and a column temperature of 30 °C were used. The drugs were monitored at 380 nm, and the injection volume was 10 μL.

2.3. CPME Procedure

The collection of the samples was performed by healthy volunteers that did not take any AMB medication. All volunteers provided their written consent after being fully informed of the procedures followed in these experiments. Serum samples were stored at −20 °C. Prior to the CPME procedure, 25 μL of each ISTD solution, 25 μL of analytes working solution (or Milli-Q water in the case of blank samples) and 200 μL of water were added to 250 μL of serum sample. No filtration or centrifugation steps were required.

For the activation of the sol–gel sorbent, the microextraction capsule was immersed in 2 mL of MeOH for 5 min, followed by immersion in 2 mL of water. Accordingly, the capsule was immersed in the sample solution (500 μL) for the adsorption of the analyte. AMB was extracted within 25 min under stirring at 500 rpm. After this step, the capsule was rinsed with H_2O and dried with a lint-free tissue. For the elution of AMB, the capsule was placed in an Eppendorf tube, and 250 μL of MeOH was added. The tube was vortexed for 10 s, and elution was completed after 5 min. After each extraction, the capsule was washed with the 2 mL of MeOH used in the activation step, left to dry, and then stored.

2.4. Multivariate Optimization

Response surface methodology associated with the central composite design was employed to highlight the optimized conditions for the sample volume, extraction time and stirring rate parameters. This procedure has been frequently used for the optimization of microextraction techniques, and it combines two-level full or fractional factorial designs with additional axial or star points and at least one point at the center of the experimental region being studied [19,20]. TIBCO Statistica software v. 13.3.0 (TIBCO Software Inc., Palo Alto, CA, USA) was employed for the experimental designs.

2.5. Method Validation

The proposed bioanalytical method was validated according to the Food and Drug Administration (FDA) as per selectivity, linearity, precision, accuracy, limit of detection (LOD) and limit of quantitation (LOQ) [21]. The selectivity was assessed by analyzing different drug-free and spiked serum samples in order to demonstrate the lack of interference from endogenous compounds. The linearity of the method was determined by plotting the ratio of the peak area of the analyte to ISTD against the nominal concentrations of matrix-matched calibration standards. The experimental data were fitted to linear regression analysis without a weighting factor. The precision and accuracy were investigated at three concentration levels on the same working day and on three consecutive days. The studied concentration levels represented the lower limit of quantitation (LLOQ), a medium quality control (MQC) level, and a high quality control (HQC) level. The acceptance criteria for the precision is that the% RSD should be less than 15% and 20% for QC levels and LLOQ, respectively. In terms of accuracy, the% relative recoveries (% RR) should be 85–115% and 80–120% for the QC and LLOQ levels, respectively. The LOD and LOQ were estimated using the $S/N = 3$ and $S/N = 10$ criteria, respectively. The AMB was considered stable for the stability test if the% RSD of the replicates was not more than 15% and the RR (%) ranged between 85–115%.

3. Results and Discussion

3.1. Optimization of CPME Parameters

The type of CPME sorbent, sample pH, elution solvent and its volume, elution time and salt concentration were investigated using the one-factor-at-a-time (OFAT) approach to achieve optimal extraction efficiency. For this purpose, each parameter was individually examined, and its optimum value was found and used for further experiments. The initial sample preparation conditions were the following, sample volume: 1 mL, type of eluent: MeOH, eluent volume: 1 mL, extraction time: 30 min, stirring rate: 400 rpm. Following the examination of these factors, response surface methodology was used during the optimization of sample volume, extraction time and stirring rate. A standard solution containing AMB at a concentration of 1.0 μg mL^{-1} was used throughout the optimization study. For each experiment, the extraction recovery (%ER) of AMB was calculated.

3.1.1. Study of CPME Sorbent and Sample pH

The initial parameter that was optimized was the type of monolithic CPME sorbent using microextraction capsules of 1 cm in length, since the volume of biological fluid is rather limited. On this basis, five different types of microextraction capsules with different polarities were examined (see supplementary material). AMB is a macrocyclic lactone that contains a carboxylic acid group (pK_a 5.7) and a basic mycosamine sugar (pK_a 10) [22]. The above monolithic sorbents were evaluated by using a standard aqueous solution of AMB at pH 7. As can be seen in Figure 1A, moderate extraction recoveries were obtained for AMB for the CPME tested. However, the polar monolithic-based sorbent CW 20M provided higher extraction efficiencies compared to the medium polarity PTHF and the non-polar sorbents (i.e., C_{18}, PDMDPS, PDMS), and thus this membrane was selected for further experiments. Based on these findings, the predominant interactions between the analyte and the polyethylene glycol sorbent may be attributed to hydrogen bonding. Moreover, at neutral pH media, weak hydrophilic and physical adsorption between AMB and the substantiate portion of the monolithic silica backbone can take place.

Another parameter included the study of the%ER of the analyte at two pH values (i.e., 3 and 7). Almost similar extraction efficiency was recorded at both studied pH values indicating that the protonated form of AMB did not significantly affect its interactions with the CW 20M sorbent. Taking into consideration the above-mentioned issue and sample handling simplicity, the neutral pH value (unadjusted pH) was selected and adopted for further experiments.

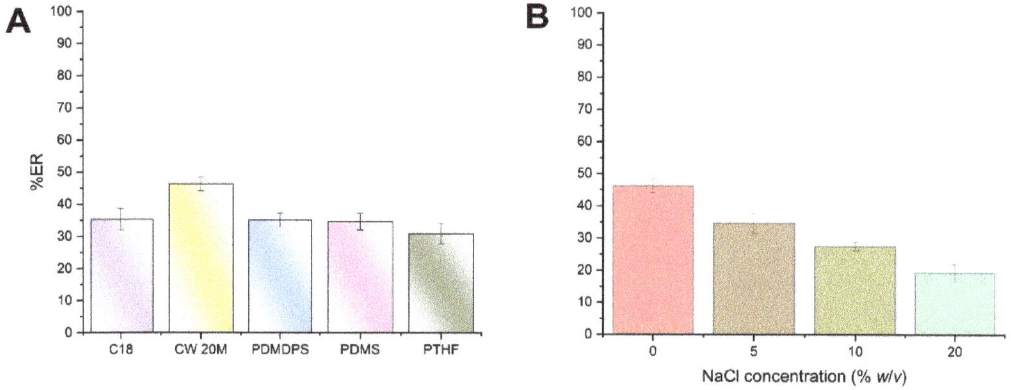

Figure 1. Effect of the (**A**) CPME sorbent type and (**B**) ionic strength (expressed as NaCl concentration,% *w*/*v*) on the ER% of AMB.

3.1.2. Study of Ionic Strength

The addition of salt can decrease the solubility of compounds with intermediate polarity, resulting in their enhanced transference to the sorbent. This phenomenon is known as the salting-out effect. However, the addition of salt can also result in an enhancement of the solution's viscosity, and thus it can diminish the salting-out effect and reduce the mass transfer. The effect of the salt addition on the ER% of AMB was assessed under variable concentrations of NaCl (i.e., 0–20% *w*/*v*). As shown in Figure 1B, the ER% values of AMB were progressively decreased by enhancing sample ionic strength. As a result, no addition of salt was chosen for the CPME protocol.

3.1.3. Study of Sample Volume, Extraction Time, and Stirring Rate

The next step included the optimization of the three parameters, namely, sample volume (V_s), extraction time ($t_{ext.}$) and stirring rate using response surface methodology through the face-centered central composite design (FC-CCD). For the experimental design, a set of 20 experiments were carried out, and they included six center point runs. In this set of experiments, the pH and the ionic strength of the sample were 7 and 0% *w*/*v*, respectively. Table 1 shows the factorial design points. The CPME experiments were randomized, aiming to eliminate any potential systematic errors. A fitted, second-order polynomial quadratic model was constructed using multivariate regression analysis.

In Table S1, the most important effects and interactions are presented as they were evaluated by ANOVA. The calculated models' R^2 were >0.9225 showing that the predicted models explain the responses adequately. The *p*-value of the "lack-of-fit" (LOF) was non-significant since it was higher than 0.05. Figure S1 shows the diagnostic plots (i.e., the plot of residuals against the predicted values and the normal probability plot of residuals). A good correlation between the actual and the predicted responses can be observed, since all data were monotonously dispersed around the line. The contour plots of AMB and a set of response surfaces are depicted in Figure 2.

Derringer's desirability function (D) was used (scaled between 0 to 1) as a fully desirable response, aiming to find the optimal CPME conditions. The desirability graphs and the prediction profiles are shown in Figure 3. The optimum conditions were 500 μL for V_s, 25 min for $t_{ext.}$, and 500 rpm for the stirring rate. Finally, six repetitive extractions were carried out under optimum conditions for their confirmation. The differences between the predicted values and the obtained values from the set of experiments were <7%.

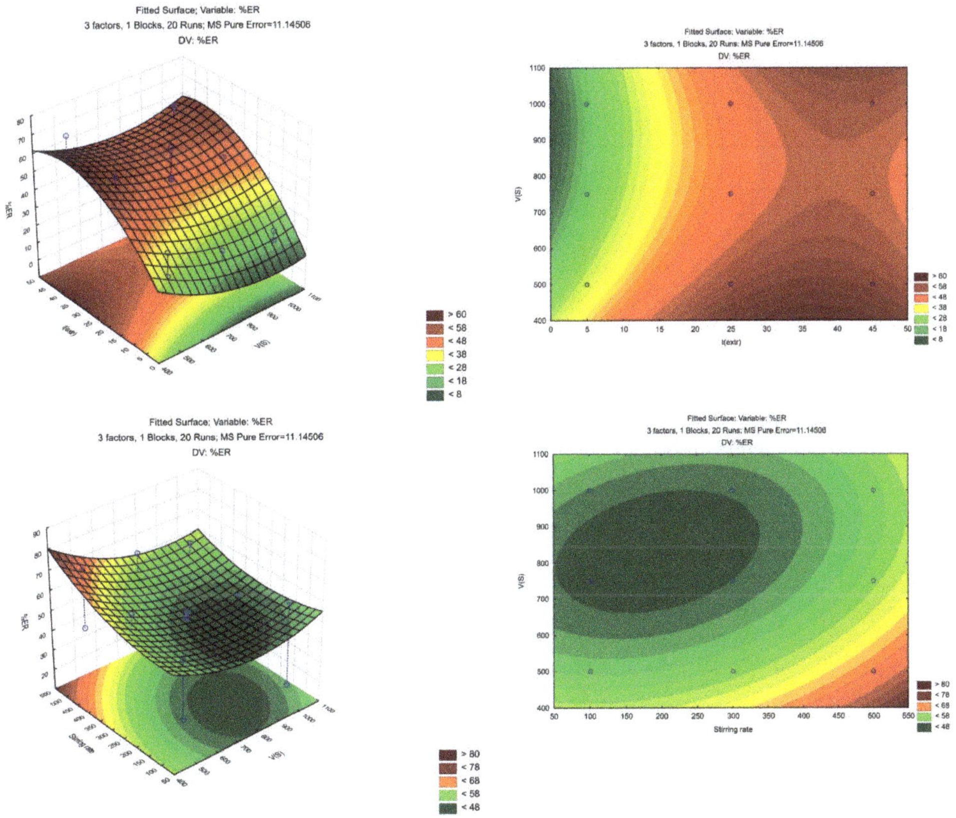

Figure 2. 3D and contour plots illustrating the effects of V_S, $t_{ext.}$, and stirring rate on the%ER of AMB.

Table 1. Experimental FC-CCD domain for the optimization of CPME parameters.

Standard Run	V_s (μL)	$t_{ext.}$ (min)	Stirring Rate (rpm)	%ER
7	1000	45	100	60.84
15 (C [1])	750	25	300	44.50
6	1000	5	500	24.71
11	750	5	300	27.51
4	500	45	500	68.65
14	750	25	500	66.43
16 (C)	750	25	300	49.49
5	1000	5	100	19.38
1	500	5	100	27.60
13	750	25	100	44.44
20 (C)	750	25	300	49.06
18 (C)	750	25	300	45.58
19 (C)	750	25	300	46.29
10	1000	25	300	48.74
3	500	45	100	57.08
2	500	5	500	40.29
8	1000	45	500	60.52
9	500	25	300	63.03
12	750	45	300	52.85
17 (C)	750	25	300	40.36

[1] C: center point.

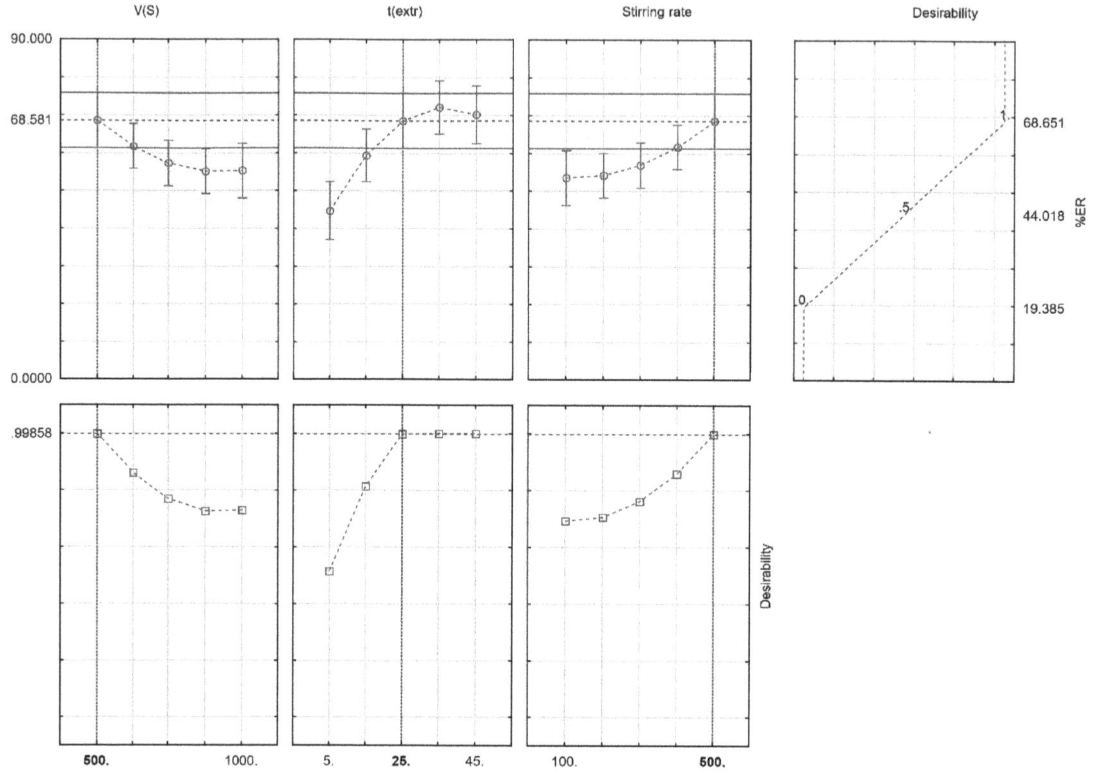

Figure 3. Profiles for predicted values and desirability function (red-colored lines designate the optimum value).

3.1.4. Study of the Elution Solvent, Volume and Elution Time

To find the optimum desorption conditions of AMB from the CPME device different parameters, including the kind of elution solvent, its volume, and the elution time were examined. The goal of this set of experiments was to provide the highest possible desorption efficiency and to avoid any undesirable carry-over phenomena.

Four eluents (i.e., MeOH, ACN, MeOH: ACN, 50:50 v/v, and 1% v/v formic acid in MeOH) with different polarities were investigated, and the results of the study are illustrated in Figure 4A. As can be seen, the elution efficiency was limited when ACN was utilized, while the usage of MeOH resulted in the highest recoveries. The 1% formic acid in MeOH did not improve the desorption performance of AMB in comparison with MeOH. Thus, MeOH was chosen as eluent. It has to be highlighted that the selection of MeOH complies with the Pfizer solvent selection guide, taking into consideration solvent toxicity [23].

Subsequently, the desorption time of the CPME procedure was studied between 2 and 20 min to guarantee sufficient contact time between MeOH and AMB. Generally, the desorption time must be as low as possible to obtain rapid protocols with high sample throughput. As shown in Figure 4B, the complete elution of AMB was achieved within 5 min. Therefore, this time span was used in further experiments.

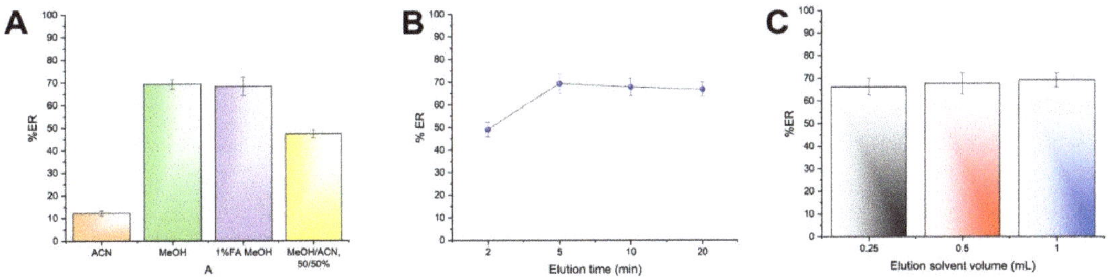

Figure 4. Effect of the (**A**) elution solvent type, (**B**) elution time, and (**C**) elution solvent volume on the%ER of AMB.

As the last step, the volume of MeOH was studied to obtain quantitative elution of AMB, as well as reduced solvent consumption in accordance with the principles of GAC [15]. Different volumes between 250–1000 μL were investigated. Figure 4C shows that efficient desorption of AMB can be achieved by 250 μL of MeOH. Thus, an organic solvent volume of 250 μL was chosen.

3.2. Method Validation

The proposed CPME-HPLC-UV protocol for the determination of AMB in human serum matrixes was validated in terms of linearity, selectivity, LOD, LOQ, precision and accuracy.

Human serum samples spiked with AMB at seven different concentration levels, between 0.10 to 10.0 μg mL^{-1}, were prepared. All samples were subjected to CPME-HPLC-UV analysis, and the peak area of AMB and ISTD were recorded. The unweighted regression equation was $y = 0.5005x - 0.0127$. The r^2 value was greater than 0.9965, showing good linearity within the examined range. The LLOQ was set at 0.10 μg mL^{-1}, and the LOD was estimated to be 0.03 μg mL^{-1}, based on the $S/N = 10$ criterion and $S/N = 3$ criterion, respectively.

Method selectivity was investigated by analyzing six different blank human serum samples. The absence of interfering peaks on the retention time of AMB and the ISTD demonstrates the selectivity of the proposed CPME-HPLC-UV method. In Figure 5, representative chromatograms from the analysis of blank and spiked samples are shown.

The intra-day method repeatability and the inter-day method accuracy and precision were examined by performing five individual extractions ($n = 5$) on the same day and by performing three extractions daily for three consecutive days ($n = 3 \times 4$) at three different concentration levels (i.e., 0.1, 1.0 and 10.0 μg mL^{-1}). The RSD values were between 4.8–12.4% for intra-day repeatability and between 3.2–9.7% for inter-day precision.

Finally, the method's intra-day and inter-day accuracy were expressed as relative recovery%, and they were evaluated by performing the different extractions that were mentioned above. All relative recoveries were acceptable, ranging between 87.0 and 113.0%, demonstrating good method accuracy.

3.3. Microextraction Capsule Reusability

The reusability of the sol–gel CW 20M microextraction capsule was studied in continuous extraction cycles. For these experiments, the same capsule was used for the extraction of AMB from a spiked serum sample at a level of 1.0 μg mL^{-1}. A criterion of 10% of extraction performance loss was used to evaluate the reusability. The results are shown in Figure 6. After the completion of 10 cycles, the loss of the extraction efficiency was not significant, demonstrating that the microextraction capsule can be used at least 10 times.

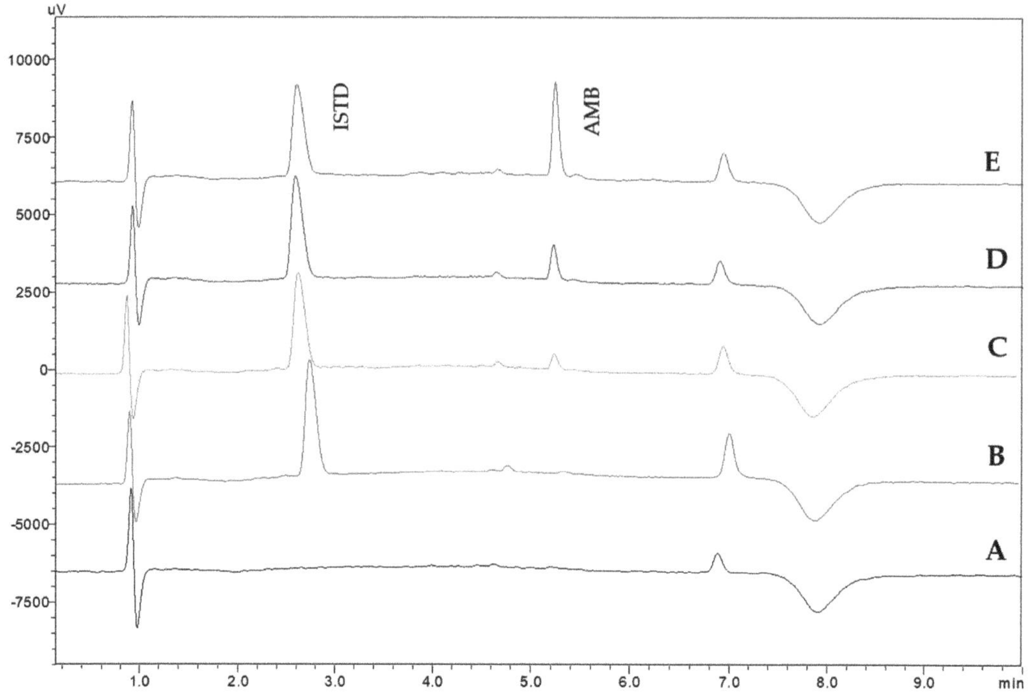

Figure 5. Representative chromatograms of the analysis of (A) blank serum and spiked with (B) ISTD, (C) ISTD and AMB (0.25 μg mL^{-1}), (D) ISTD and AMB (0.50 μg mL^{-1}), and (E) ISTD and AMB (1.0 μg mL^{-1}) using the proposed CPME-HPLC-UV method. c(ISTD): 20 μg mL^{-1}.

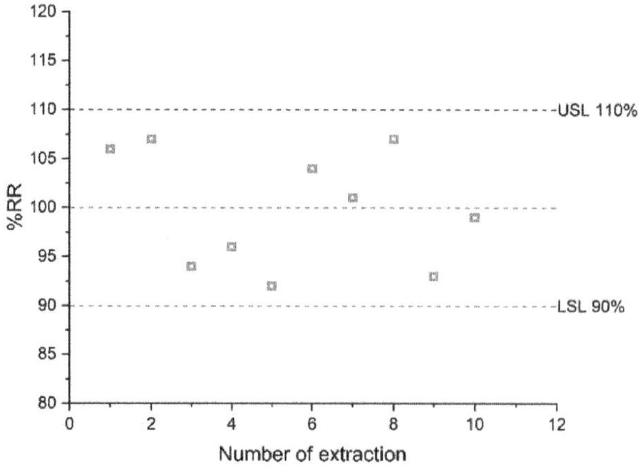

Figure 6. Reusability of CW 20M CPME membrane.

3.4. Sample Stability

The stability of the target analyte in the serum sample was examined at 0, 4, and 8 h, stored at 25 °C and at 0, 8, and 24 h stored at +4 °C. For this purpose, blank human serum samples were spiked with AMB at a concentration of 1.0 μg mL^{-1}, and they were stored under the studied conditions. After this time span, the samples were subjected to

CPME-HPLC-UV analysis. A criterion of ±15% as a deviation from the nominal value was set to assess the stability. In accordance with previous studies, no degradation of AMB was observed under these conditions.

3.5. Application to Human Serum Samples

In order to prove the applicability of the CPME-HPLC-UV protocol, different individual human serum samples were analyzed. For this purpose, the authentic samples were spiked at three different concentration levels (i.e., 0.1, 1.0, and 10.0 μg mL^{-1}) and they were subjected to the developed method. As shown in Table 2, the relative recovery was 83.9–114.6%, while the%RSD was 4.3–9.8%, showing good method applicability.

Table 2. Relative recoveries of AMB in human serum samples.

Added Concentration (μg mL^{-1})	Relative Recoveries (%) (RSD%, $n = 3$)		
	Sample 1	**Sample 2**	**Sample 3**
0.10	86.4 (5.0)	88.4 (4.3)	83.9 (8.7)
1.0	114.6 (6.5)	108.7 (5.9)	98.7 (9.8)
10.0	95.2 (4.4)	95.3 (7.2)	94.7 (3.9)

3.6. Evaluation of the Green Character of the Proposed Method by ComplexGAPI Index

For the complete evaluation of the herein proposed method, the complexGAPI index [24] was used to assess its green character. As such, the environmental friendliness of the CPME-HPLC-UV protocol was examined from the collection of the sample to the final step of the analytical determination [25], while the synthetical route for the microextraction capsules was also taken into consideration. Figure 7 shows the obtained pictogram for the proposed method. The preparation of the microextraction devices meets most of the requirements of the index (green color), resulting in an environmentally friendly procedure. Regarding the sample collection/sample preparation and analytical determination of AMB, many of the criteria are also met (e.g., low organic solvent consumption and reduced waste generation) in terms of the green elements of the final protocol.

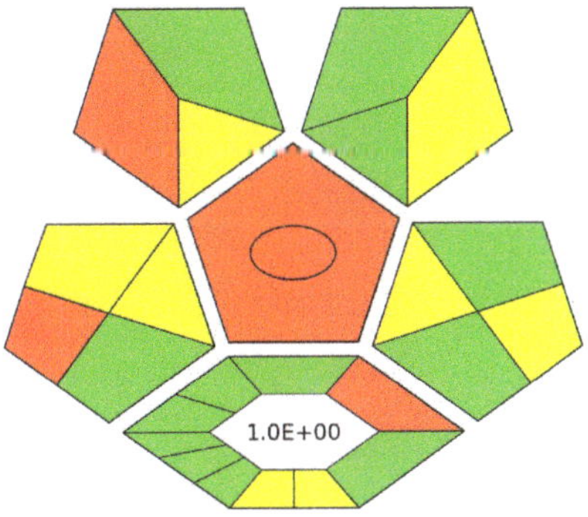

Figure 7. ComplexGAPI pictogram for the CPME-HPLC-UV protocol for the determination of AMB in human serum.

4. Conclusions

CPME serves as a useful bioanalytical technique that efficiently simplifies the sample preparation of complex biological fluids by integrating the stirring and filtration mechanism in one capsule. The utilization of this technique results in high sample throughput, good extraction performance and compliance with many of the principles of GAC. Under optimum adsorption and desorption conditions, good figures of merit were obtained. The sol–gel CW 20M microextraction capsules were found to be reusable for at least 10 times. In the final step, the proposed methodology was successfully applied for the determination of AMB in real human serum samples obtained from different volunteers.

Supplementary Materials: The following supporting information can be downloaded at: https://www.mdpi.com/article/10.3390/separations9120433/s1. Preparation of capsule device, Figure S1: Plot of observed and predicted values for FC-CCD for AMB.; Table S1: Analysis of variance (ANOVA) for FC-CCD for ER% of AMB.

Author Contributions: Conceptualization, C.K.Z.; methodology, C.K.Z., A.K. (Abuzar Kabir) and N.M.; software, C.K.Z.; validation, N.M.; formal analysis, N.M., A.K. (Anastasia Korpeti) and A.K. (Abuzar Kabir); investigation, N.M., A.K. (Anastasia Korpeti) and A.K. (Abuzar Kabir); data curation, N.M. and A.K. (Anastasia Korpeti); writing—original draft preparation, C.K.Z. and N.M.; writing—review and editing, C.K.Z., A.K. (Abuzar Kabir) and K.G.F.; supervision, C.K.Z.; project administration, C.K.Z. All authors have read and agreed to the published version of the manuscript.

Funding: This research received no external funding.

Informed Consent Statement: Written informed consent has been obtained from the volunteers to publish this paper.

Data Availability Statement: Data is contained within the article or supplementary material.

Conflicts of Interest: The authors declare no conflict of interest.

References

1. Espada, R.; Josa, J.M.; Valdespina, S.; Dea, M.A.; Ballesteros, M.P.; Alunda, J.M.; Torrado, J.J. HPLC Assay for Determination of Amphotericin B in Biological Samples. *Biomed. Chromatogr.* **2008**, *22*, 402–407. [CrossRef] [PubMed]
2. Barrett, J.P.; Vardulaki, K.A.; Conlon, C.; Cooke, J.; Daza-Ramirez, P.; Evans, E.G.V.; Hawkey, P.M.; Herbrecht, R.; Marks, D.I.; Moraleda, J.M.; et al. A Systematic Review of the Antifungal Effectiveness and Tolerability of Amphotericin B Formulations. *Clin. Ther.* **2003**, *25*, 1295–1320. [CrossRef] [PubMed]
3. Balabathula, P.; Janagam, D.R.; Mittal, N.K.; Mandal, B.; Thoma, L.A.; Wood, G.C. Rapid Quantitative Evaluation of Amphotericin B in Human Plasma, by Validated HPLC Method. *J. Bioequivalence Bioavailab.* **2013**, *5*, 121–124. [CrossRef]
4. He, Y.; Concheiro-Guisan, M. Microextraction Sample Preparation Techniques in Forensic Analytical Toxicology. *Biomed. Chromatogr.* **2019**, *33*, e4444. [CrossRef] [PubMed]
5. Manousi, N.; Samanidou, V. Green Sample Preparation of Alternative Biosamples in Forensic Toxicology. *Sustain. Chem. Pharm.* **2021**, *20*, 100388. [CrossRef]
6. Samanidou, V.F.; Metaxa, A.S.; Papadoyannis, I.N. Direct Simultaneous Determination of Uremic toxins: Creatine, Creatinine, Uric acid, and Xanthine in Human Biofluids by HPLC. *J. Liq. Chromatogr. Relat. Technol.* **2002**, *25*, 43–57. [CrossRef]
7. Gałuszka, A.; Migaszewski, Z.; Namieśnik, J. The 12 Principles of Green Analytical Chemistry and the SIGNIFICANCE Mnemonic of Green Analytical Practices. *TrAC—Trends Anal. Chem.* **2013**, *50*, 78–84. [CrossRef]
8. Silva, C.; Cavaco, C.; Perestrelo, R.; Pereira, J.; Câmara, J.S. Microextraction by Packed Sorbent (MEPS) and Solid-Phase Microextraction (SPME) as Sample Preparation Procedures for the Metabolomic Profiling of Urine. *Metabolites* **2014**, *4*, 71–97. [CrossRef]
9. Carasek, E.; Morés, L.; Huelsmann, R.D. Disposable Pipette Extraction: A Critical Review of Concepts, Applications, and Directions. *Anal Chim Acta* **2022**, *1192*, 339383. [CrossRef]
10. Pedersen-bjergaard, S.; Rasmussen, K.E. Bioanalysis of Drugs by Liquid-Phase Microextraction Coupled to Separation Techniques. *J. Chromatogr. B* **2005**, *817*, 3–12. [CrossRef]
11. Tienpont, B.; David, F.; Desmet, K.; Sandra, P. Stir Bar Sorptive Extraction-Thermal Desorption-Capillary GC-MS Applied to Biological Fluids. *Anal. Bioanal. Chem.* **2002**, *373*, 46–55. [CrossRef] [PubMed]
12. Tartaglia, A.; Covone, S.; Rosato, E.; Bonelli, M.; Savini, F.; Furton, K.G.; Gazioglu, I.; D'Ovidio, C.; Kabir, A.; Locatelli, M. Fabric Phase Sorptive Extraction (FPSE) as an Efficient Sample Preparation Platform for the Extraction of Antidepressant Drugs from Biological Fluids. *Adv. Sample Prep.* **2022**, *3*, 100022. [CrossRef]
13. Manousi, N.; Kabir, A.; Furton, K.G.; Samanidou, V.F.; Zacharis, C.K. Exploiting the Capsule Phase Microextraction Features in Bioanalysis: Extraction of Ibuprofen from Urine Samples. *Microchem. J.* **2021**, *172*, 106934. [CrossRef]
14. US10352833B2—Microextraction Capsules and Method of Making—Google Patents. Available online: https://patents.google.com/patent/US10352833B2/en (accessed on 5 December 2022).
15. Janicka, P.; Płotka-Wasylka, J.; Jatkowska, N.; Chabowska, A.; Fares, M.Y.; Andruch, V.; Kaykhaii, M.; Gębicki, J. Trends in the New Generation of Green Solvents in Extraction Processes. *Curr. Opin. Green Sustain. Chem.* **2022**, *37*, 100670. [CrossRef]
16. López-Lorente, Á.I.; Pena-Pereira, F.; Pedersen-Bjergaard, S.; Zuin, V.G.; Ozkan, S.A.; Psillakis, E. The Ten Principles of Green Sample Preparation. *TrAC—Trends Anal. Chem.* **2022**, *148*, 116530. [CrossRef]

17. Mamounas, G.; Manousi, N.; Kabir, A.; Furton, K.G.; Mystridis, G.A.; Vizirianakis, I.S.; Zacharis, C.K. Designing an "All-in-One" Microextraction Capsule Device for the Liquid Chromatographic-Fluorescence Determination of Doxorubicin and Its Metabolites in Rat Plasma. *J. Chromatogr. A* **2022**, *1680*, 463432. [CrossRef] [PubMed]
18. Manousi, N.; Kabir, A.; Furton, K.G.; Tzanavaras, P.D.; Zacharis, C.K. In Situ Synthesis of Monolithic Sol–Gel Polyethylene Glycol-Based Sorbent Encapsulated in Porous Polypropylene Microextraction Capsules and Its Application for Selective Extraction of Antifungal and Anthelmintic Drugs from Human Urine. *Microchem. J.* **2022**, *180*, 107594. [CrossRef]
19. Ferreira, S.L.C.; Bruns, R.E.; da Silva, E.G.P.; dos Santos, W.N.L.; Quintella, C.M.; David, J.M.; de Andrade, J.B.; Breitkreitz, M.C.; Jardim, I.C.S.F.; Neto, B.B. Statistical Designs and Response Surface Techniques for the Optimization of Chromatographic Systems. *J. Chromatogr. A* **2007**, *1158*, 2–14. [CrossRef]
20. Manousi, N.; Vlachaki, A.; Kika, F.S.; Markopoulou, C.K.; Tzanavaras, P.D.; Zacharis, C.K. Salting-out Homogeneous Liquid-Liquid Microextraction for the Determination of Azole Drugs in Human Urine: Validation Using Total Error Concept. *J. Sep. Sci.* **2022**, *45*, 1240–1251. [CrossRef]
21. FDA, Bioanalytical Method Validation Guidance, Food Drug Adm. 2018. Available online: https://www.Fda.Gov/Regulatory-Information/Search-Fda-Guidance-Documents/BioanalytiCal-Method-Validation-Guidance-Industry (accessed on 10 December 2022).
22. Faustino, C.; Pinheiro, L. Lipid Systems for the Delivery of Amphotericin B in Antifungal Therapy. *Pharmaceutics* **2020**, *12*, 29. [CrossRef]
23. Byrne, F.P.; Jin, S.; Paggiola, G.; Petchey, T.H.M.; Clark, J.H.; Farmer, T.J.; Hunt, A.J.; Mcelroy, C.R.; Sherwood, J. Tools and Techniques for Solvent Selection: Green Solvent Selection Guides. *Sustain. Chem. Process.* **2016**, *4*, 7. [CrossRef]
24. Płotka-Wasylka, J.; Wojnowski, W. Complementary Green Analytical Procedure Index (ComplexGAPI) and Software. *Green Chem.* **2021**, *23*, 8657–8665. [CrossRef]
25. Płotka-Wasylka, J. A New Tool for the Evaluation of the Analytical Procedure: Green Analytical Procedure Index. *Talanta* **2018**, *181*, 204–209. [CrossRef] [PubMed]

separations

MDPI

Article

Hyphenation of Thermodesorption into GC × GC-TOFMS for Odorous Molecule Detection in Car Materials: Column Sets and Adaptation of Second Column Dimensions to TD Pressure Constraints

Romain Klein [1,2,*], José Dugay [1], Jérôme Vial [1], Didier Thiébaut [1,*], Guy Colombet [2], Donatien Barreteau [2] and Guillaume Gruntz [2]

[1] Laboratoire Sciences Analytiques Bioanalytiques et Miniaturisation, CBI, ESPCI Paris, Université PSL, CNRS, 75005 Paris, France; jose.dugay@espci.fr (J.D.); jerome.vial@espci.fr (J.V.)

[2] Renault Group, Materials Engineering Department, 1 Avenue du Golf, 78084 Guyancourt, France; guy.colombet@renault.com (G.C.); guillaume.gruntz@renault.com (G.G.)

* Correspondence: romain.klein@espci.fr (R.K.); didier.thiebaut@espci.fr (D.T.)

Abstract: Vehicle interior air quality is an issue of growing interest among car manufacturers and customers. GC-MS is the benchmark method for the analysis of indoor air or material emissions. It is suitable for the quantification of target pollutants and the most abundant compounds. It fails, however, to uncover the true molecular complexity of these samples. In the present study, we describe the development of a TD-GC × GC-TOFMS method designed to detect polar and potentially odorous molecules in car material emissions. Attention is paid to the hyphenation of the thermodesorber and the gas chromatograph, both at software and hardware levels, and the constraints due to pressure limitations on the thermodesorber (evaluated at 414 kPa/60 psi at the end of the temperature ramp and at 138 kPa/20 psi at rest). A compromise was made for the 2D column length and diameter to balance separation and pressure (50 × 0.18 × 0.18 cm × mm × µm + 60 cm transfer line selected). On various materials, we were able to observe several hundreds of polar molecules, among them were between 75 and 150 odorants per material. This work lays the foundation for the widespread screening of potential odorants in car material emissions.

Keywords: GC × GC; VIAQ; thermodesorption; air quality

1. Introduction

Modern cars are very complex constructs made of a wide variety of materials, of which a growing proportion is synthetic. Most of these materials are likely to emit volatile organic compounds (VOCs) in a vehicle's cabin, constituting a potential nuisance to the drivers and passengers that needs to be monitored [1]. For this reason, several studies relying on gas chromatography (GC) have already been conducted on the molecules present in the car cabins of newly produced cars [2–4] or used ones [5,6] or to evaluate the effect of influencing factors (such as vehicle price or driving and parking behaviour) [7–9]. The vast majority of the molecules identified in these studies were aliphatic or aromatic hydrocarbons, and their wide diversity often resulted in low-resolution regions in the chromatograms wherein not all peaks could be identified [5,6].

In addition to monitoring vehicle interior air quality (VIAQ), car manufacturers have also developed methods for assessing the odour of these emissions [10,11], such as the series of VDA 276 (Verband der Automobilindustrie, from the combined work of German car manufacturers) [12] and ISO 12219 standards [13]. From this experience, it has been noted that minor changes in the formulation of a material may have a strong impact on the perceived odour of the emissions. Indeed, global odour often results from polar molecules representing a minority in terms of global abundance for this category of materials [14], and a mixture of several odorants can result in an odour very different from the original odorants smelled separately [15,16]. This effect can be emphasised for recycled elements, as the raw materials they are made from inherently have a variable composition, yielding more potentially odorous molecules.

To detect and prevent the presence of malodorous molecules, car manufacturers need analytical tools. Methods relying on human smelling capacity, such as panels [17–21] or GC–olfactometry (GC-O) techniques [22–24], exist, but they are very expensive and time-consuming. Panels give no information on molecular composition, and GC-O requires molecules to be separated on the chromatogram so that they can be identified. However, as previously mentioned, typical GC methods fail to fully separate the complex samples that are material emissions. Moreover, odorous polar molecules are rarely detected and are not separated properly from the barely to non-odorous major non-polar compounds in common GC analyses [5,14].

Thus, it is necessary to develop new methods to better detect potentially odorous polar molecules in vehicle interior air. Comprehensive bidimensional gas chromatography (GC × GC), combining two different columns interfaced by a modulator, is a modern analytical technique offering three main improvements in comparison to classical GC: (i) a much higher peak capacity [25], (ii) a structured 2D chromatogram for chemical family identification [26], and (iii) better sensitivity due to refocusing on the modulator [27]. All these elements give GC × GC the potential to separate low-concentration potentially odorous polar molecules from the large number of hydrocarbons present in car material emissions.

In this paper, we present a method combining thermodesorption (TD), comprehensive bidimensional gas chromatography (GC × GC), and time-of-flight mass spectrometry (TOFMS) to undercover the true molecular complexity of material emissions. We aim to create a database of odorants from various car material emissions, focusing on individual molecules instead of clusters belonging to the same family. The present work deals with analytical instrument optimisation and early results towards odorous molecules.

To this day, no tool reliably predicts whether or not a molecule is odorous from its structure alone [21,28,29]. Therefore, any polar molecule may be considered an analyte of interest, and most chemical families must be compatible with analytical conditions. For this broad untargeted screening, the analytical method developed was obtained through an experimental process, with a final goal of observing the largest variety of polar molecules. The present work does not aspire to explain odours by the individually detected odorants alone but to confront, in future works, the global chemical composition of emissions with their global odour perceived by humans.

In the first part, we detail the implementation of our method for the selection of the column set in terms of stationary phase combination. In our instrumental configuration, detailed in the Section 2, the TD was subjected to the total column back pressure at rest and during the whole temperature program. To the best of our knowledge, few experimental TD-GC × GC studies have included aspects on pressure drop [30,31], and none of them have reported compatibility issues between high pressure and TD. As we faced such issues, we provide a discussion in the second part regarding the experiments conducted to adapt the system, through ^{2}D column dimension variations, while trying to maintain the separation performance of the column set selected in the first part.

2. Materials and Methods

2.1. Sample Preparation

All material samples were provided by Renault Group (Guyancourt, France). A wide variety of materials, chosen to be representative, were used for prospective experiments. Among them were 2 ABS (Acrylonitrile Butadiene Styrene), 1 PCABS (Poly Carbonate–ABS), 4 different PP (Poly Propylene), 1 natural and 1 synthetic skin, 6 textiles, and 1 putty. They were complemented by 4 interior air samples. For the Part A experiments, one PP and one synthetic skin were selected as model samples. In Part B, we studied one trunk mat, one Polyurethane foam (PU), and one textile.

The materials were sampled using an internal Renault method adapted from ISO 12219-3 [32]. Discs of materials with a diameter of 6.4 cm were cut from a plate and placed in microchambers MCTE250T, provided by Markes International (Bridgend, UK). The thickness of these materials was variable (a few mm), but the exposed surface was constant at 32 cm^2. The materials were heated for 20 min at 65 °C at a dew point of 10.4 °C. Under a stream of pure air at 50 mL/min, 1 L of their emissions was collected in Tenax TA tubes (Poly(2,6-diphenylphenylene oxide)), supplied by Sigma-Aldrich (Saint-Quentin-Fallavier, France). The tubes were stored refrigerated and analysed at most two weeks after sampling.

As samples of material emissions were available in limited amounts, system suitability tests were conducted using a mixture of 48 compounds covering a large number of chemical families (see Figure S1 for composition and general peak positions). These samples were prepared by depositing a droplet of standard at the top of a sampling tube, assisted by a gas stream, using the Calibration Solution Loading Rig (CSLR) from Markes International (Bridgend, UK). This method previously achieved a reliable preparation of standard tubes, as well as quantitation for these 48 compounds for a 10 ng deposit, with a limit of detection of the order of 0.1–1 ng [33].

2.2. TD-GC Hyphenation

Once loaded with material emissions, the tubes were analysed using a system composed of a TD100-xr from Markes International (Bridgend, UK) and a GC × GC-TOFMS Pegasus BT4D from LECO (Villepinte, France). The first device was monitored using the software Markes Instrument Control version 2.0.5 and the second one by ChromaTOF version 5.51. The Pegasus BT4D combines a TOFMS detector and a GC device equipped with a modulator and a secondary oven for a potential secondary GC column, thus making it suitable for GC × GC. From the thermodesorber perspective, the association is the same in the GC or GC × GC configuration. According to both instrument manufacturers, this was the first time they had ever been combined.

The global system was supplied with helium as a carrier gas (Alphagaz 1, Air Liquide, France), plugged into both instruments (the TD supply is used only for desorption from sample to trap, which works independently and will therefore be ignored hereafter). The GC carrier gas was pressure-regulated using a GC Electronic Pressure Controller (EPC), then sent into the TD. The TD's valves maintained a constant flow of 1.2 mL/min into the GC device at all times.

All tubes were desorbed at a flowrate of 50 mL/min and at 270 °C for 10 min (including 3 min to reach the temperature) into a cold trap "General Purpose" U-T11GPC-2S (Markes International, Bridgend, UK) at −30 °C. Then, the cold trap was desorbed at 300 °C for 5 min with a split flow of 20 mL/min and a column inlet flow of 1.2 mL/min with helium as a carrier gas towards the GC × GC through a 150 cm deactivated silica capillary connected to the ^{1}D column. Split flow was redirected (not represented in Scheme 1) towards the initial sampling tube for recollection (the tube temperature was between 50 and 60 °C).

In this instrumental configuration, the GC EPC controls the pressure for the whole system, and the TD assumes the role of the GC injector, only controlling the split flow during injection (the GC's own injector is bypassed). Moreover, 3 mL/min of carrier gas is redirected from the TD system to the EPC for pressure feedback. To notify this configuration to the GC instrument, its septum purge time during the injection must be set at 999.99 min, and the feedback flow has to be set with the "septum purge flow" parameter. These carrier gas flows are summarised in Scheme 1.

At the beginning of every analysis, the ChromaTOF software relays the run instructions to the GC device, with the risk of overwriting the previous settings and disturbing communication between the instruments. It also does not allow for the use of methods wherein the septum purge time exceeds the oven program, making only 1000 min programs possible. To solve this issue, we conceived a GC method bearing the gas-monitoring instruction (pressure-controlled constant helium flow of 1.2 mL/min) with septum purge time at 59,999.4 s, a septum purge flow of 3 mL/min, and a fake 1000 min oven program.

In later experiments, the GC programs on ChromaTOF had their injector parameter set to "none". The oven program could still be modified freely, but the septum purge time could not, preventing its overwriting. Thus, the hyphenation of the TD and GC × GC instruments was achieved at a software level.

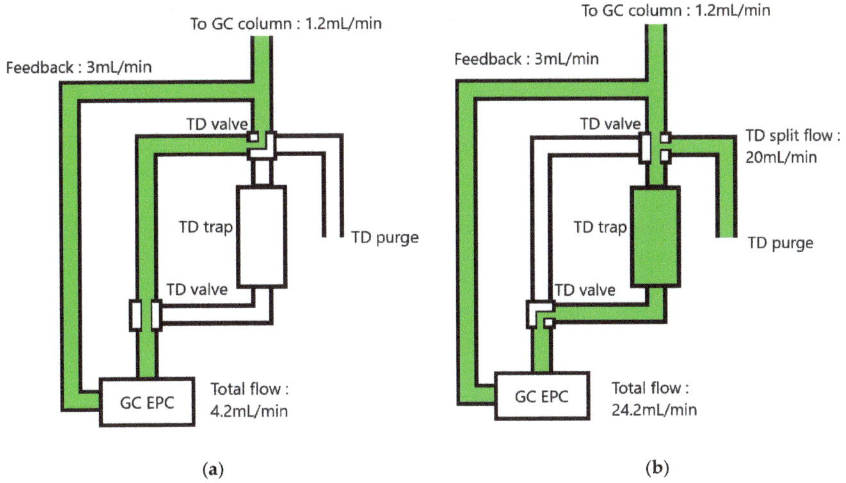

Scheme 1. A simplified scheme of the carrier gas flow in the thermodesorber (**a**) at rest/during program and (**b**) during injection. Carrier gas flows are represented in green.

2.3. Global GC × GC–TOFMS Settings

The different sets of columns used for GC × GC separations are summarised in Table 1. All the ^{1}D columns shared the same dimensions: 30 m × 0.25 mm × 0.25 μm. They were provided by Agilent (Les Ulis, France), Restek (Lisses, France) and Phenomenex (Le Pecq, France). The GC × GC was used with a temperature program starting at 40 °C, held for 2 min, and then heated up to 220 °C at 2 °C/min. The temperature of the secondary oven was set 5 °C higher than that of the primary oven, and the modulator temperature was set 15 °C higher than that of the secondary oven. The modulation period of the quad-jet modulator ranged from 2 to 12 s and was adapted to the highest ^{2}D retention time observed to prevent wrapping around. Cold jet time was set at a sixth of the modulation period during preliminary experiments and 0.5 s after.

Table 1. Stationary phases and dimensions of column sets used in the experiments in Part A; material limitations prevented the use of the same secondary column length; Rxi: Restek; ZB: Phenomenex; DB: Agilent.

Column Set	Short Name	^{1}D phase	^{2}D Dimensions (cm × mm × μm)	^{2}D Phase	Configuration	Modulation Period (s)
Rxi-5MS–ZB-WAX	5–WAX	5% diphenyl	195 × 0.1 × 0.1	100% polyethylene-glycol	Normal	12
Rxi-5MS–ZB-50	5–50	5% diphenyl	175 × 0.1 × 0.1	50% diphenyl	Normal	10
DB-17MS–DB-5	50–5	50% diphenyl	115 × 0.1 × 0.1	5% diphenyl	Reverse	6
Rxi-5MS–DB-210	5–210	5% diphenyl	118 × 0.1 × 0.1	50% trifluoropropyl	Normal	5
ZB-1701–DB-5	1701–5	14%-cyanopropyl-phenyl	84 × 0.1 × 0.1	5% diphenyl	Reverse	2
Rxi-5MS–ZB-1701	5–1701	5% diphenyl	122 × 0.1 × 0.1	14%-cyanopropyl-phenyl	Normal	6

During the Part B experiments, only the 5–1701 set was used but with various ^{2}D column dimensions. A segment of 7 cm of the ^{2}D column was present in the modulator and the 43 cm remaining in the secondary oven, while the junction with the ^{1}D column was right at the entrance of the modulator. For convenience, a 60 cm transfer line in deactivated silica matching the diameter of the secondary column was also added between the ^{2}D column and the detector (including 31 cm inside the MS transfer line at 250 °C, the rest being in the secondary oven). The temperature offsets of the secondary oven and modulator with respect to the primary oven were raised to 15 and 30 °C, respectively. The modulation period of the quad-jet modulator ranged from 2 (for 50 cm column) to 4 s (for 100 cm column), consisting of 2 cycles with 0.5 s of cold jet and 0.5 to 1.5 s of hot jet.

The mass spectrometer was used with electronic ionisation at 70 eV, using a scan range of m/z from 45 to 300 at a scan frequency of 200 Hz.

Both acquisition and data processing were conducted using ChromaTOF software. A general Peak Finding (PF) method was employed with an S/N ratio of 500. Most of the time in GC × GC, the ^{1}D peak of a compound is sampled into several ^{2}D sub-peaks over several modulation periods, resulting in a spot on the colour plot. Such sub-peaks stemming from successive modulations, with similar ^{2}D retention times and identical m/z, are combined by ChromaTOF into a single line of data (individual mass spectra and area are summed). This is what we will call a peak hereafter. Peaks in the aliphatic hydrocarbon elution region were excluded. After that, manual checking was conducted to delete artefact peaks or combine duplicated peaks of successive modulations into a single one. All peaks were attributed based on their mass spectra by comparison with the NIST 2020 database, and they received a score depending on their similarity with the closest match, ranging up to 1000. Peaks with a similarity below 700 were automatically deleted. No changes were made to the proposed attributions.

3. Results

3.1. Part A, Choice of Set of Column

3.1.1. Preliminary Experiments

A wide variety of vehicle material emissions were analysed on the column sets presented in Table 1 in order to screen the largest number of polar molecules. For all the sets, aliphatic hydrocarbons were separated from more polar molecules on our settings. Polar molecule families had variable retention times depending on the secondary column phase, especially for the 5–WAX configuration for benzaldehyde, enones, and lactones, which compelled modulation periods of up to 12 s with some wrapping around still occurring. In this configuration, these excessively large retention times were detrimental to global separation. The 5–50 configuration presented little resolution between polar non-aromatic compounds, making their identification difficult. The 5–210 configuration did not allow for the observation of as many peaks overall as the other configurations. The three remaining column configurations out of the six presented in Table 1 achieved satisfactory separations for preliminary experiments in terms of peak number and separation space occupancy. These configurations consisted of one normal and two reverse configurations: 5–1701, 1701–5, and 50–5.

During these experiments, cold jet duration was left to the default value, being a sixth of the modulation period (and half of hot jet duration). This may have had an impact on peak broadening or a too high retention of high boiling point compounds at the end of the analysis. This should not impact the previous conclusions, as the drawbacks of the rejected configurations were too high. For later experiments, cold jet duration was fixed at 0.5 s. Due to the large variety of studied samples, no further optimisation was made on this parameter to keep the method versatile. For more specific applications, it could be possible to increase cold jet duration at the beginning of the analysis to better trap light compounds and decrease it at the end to release heavy ones better.

From this first screening, two materials were selected as model samples because they covered the largest variety of molecules and situations: a PP emitting a large number of non-polar molecules and a wide variety of diluted polar ones and a synthetic leather emitting a large panel of polar molecules. Thus, these two materials alone were considered sufficient for assessing the separation quality of a column configuration.

3.1.2. Column Phases Selection

The three remaining column configurations needed to be compared using objective numerical criteria. According to our objective, i.e., observing the largest possible number of potentially odorous polar molecules, we decided to use as a performance criterion the number and quality of observed peaks based on similarity criteria regarding the NIST 2020 database. Our hypothesis was that high similarity corresponded to a low-noise and well-resolved peak, regardless of a correct attribution. From our own experience, the true nature of a peak with a similarity above 900 is often close to the proposition, belongs to the same chemical family for a similarity above 800, and may remain undetermined below 800.

The peaks in the aliphatic hydrocarbon elution region, bearing little to no odour and therefore being of no interest in our study, were excluded. Apart from them, all peaks were considered as targets. The number of peaks above each similarity threshold was counted, and the result is presented in Figure 1. The corresponding chromatograms are available in Figure S2. Both materials presented very complex emissions, with several hundreds of relevant peaks detected. These results also stress the contribution of GC × GC over classical GC, as the peak capacity in classical GC could only reach one thousand in ideal and near-theoretical conditions [25,34], meaning that classical GC can only manage the separation of one or two hundred compounds at most.

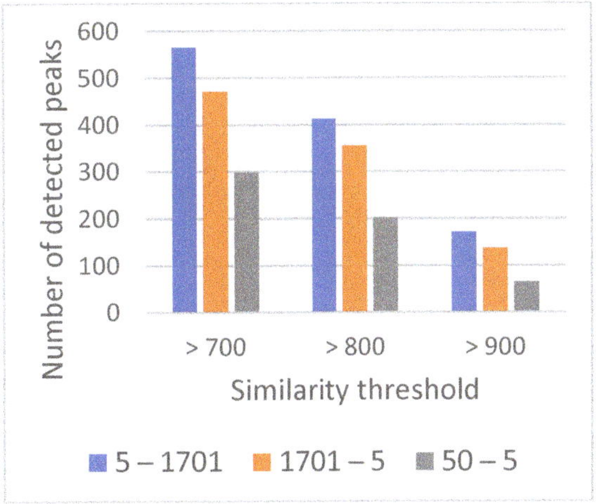

(a)

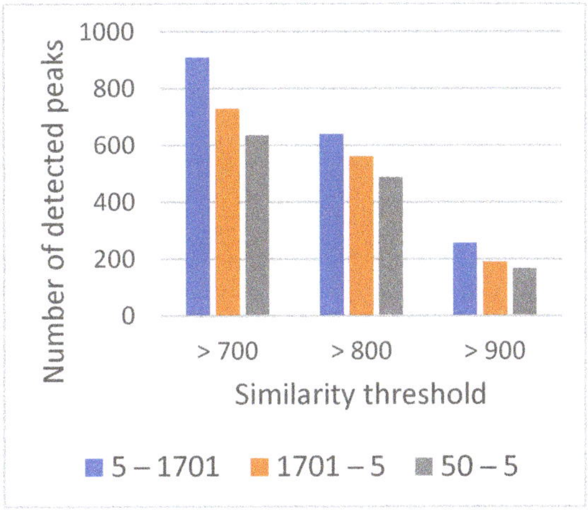

(b)

Figure 1. Number of peaks per similarity threshold for polypropylene (**a**) and synthetic leather (**b**).

The same observations could be made for both the mostly non-polar emitting PP and the mostly polar emitting leather: the 5–1701 column configuration yielded the highest number of detected peaks for every similarity threshold. Therefore, a first column with a low-polarity 5%-diphenyl phase and a second column with a polar 14%-cyanopropylphenyl phase were selected.

3.2. Part B, Selecting the Dimensions of the ^2D Column

The Influence of the Length and Diameter of the Second Dimension Column

In our instrumental configuration, the thermodesorber bypassed and assumed the role of the injector. Therefore, it had to withstand heavy pressure, because it was connected to two columns in series, for an extended time. During the experiments described in Part

A, the pressures at the inlet of the TD were generally around 207 kPa/30 psi at rest (40 °C in the primary oven) and 414 kPa/60 psi at the end of the temperature program (220 °C). According to the thermodesorber's manufacturer, this device must not operate above 60 psi, a respected condition that was adhered to. However, working regularly at such pressures increased the maintenance frequency and may prove to be detrimental for the TD in the long term. A pressure of 138 kPa/20 psi at rest seemed a safer limit, and any further decrease would benefit the device.

In GC × GC, the secondary column bears the largest part of a pressure drop. For example, a 50 cm column with an internal diameter of 0.1 mm is equivalent to a 20 m column with an internal diameter of 0.25 mm [35]. Therefore, based on the protocol of Part A and the selected column configuration (5–1701), it was decided to study the effects of the secondary column diameter and length on global pressure in order to preserve the thermodesorber. At the same time, the quality of separation already achieved needed to be maintained, a factor that was evaluated by the same means as in Part A, i.e., according to the number of potential odorants observed.

Thus, four configurations for the secondary column were studied: two lengths, 50 and 100 cm, and two combinations of diameters and stationary phase thicknesses, 0.1 × 0.1 and 0.18 × 0.18 mm × µm, to maintain constant the phase ratio of β = 500. After the second column, as a 60 cm transfer line with a matching diameter was added before the mass spectrometer, the length of the secondary column had to be shortened compared to that in Part A. Out of these four configurations, the 100 cm long and 0.18 mm wide one presented abnormally high pressure (ranging from 210 to 363 kPa) and poor separation overall. These two elements point out a probable issue with the column or its installation. Unfortunately, it was not possible to redo the experiments due to the limited number of samples; this configuration was not included in the following results.

The effect of ^{2}D configurations on pressure is presented in Figure 2. Only the 50 cm long and 0.18 mm wide column achieved a safety threshold of 138 kPa/20 psi at rest. It illustrated experimentally the much larger impact on the pressure drop of the secondary column in comparison to the primary one in bidimensional gas chromatography, despite its much smaller length.

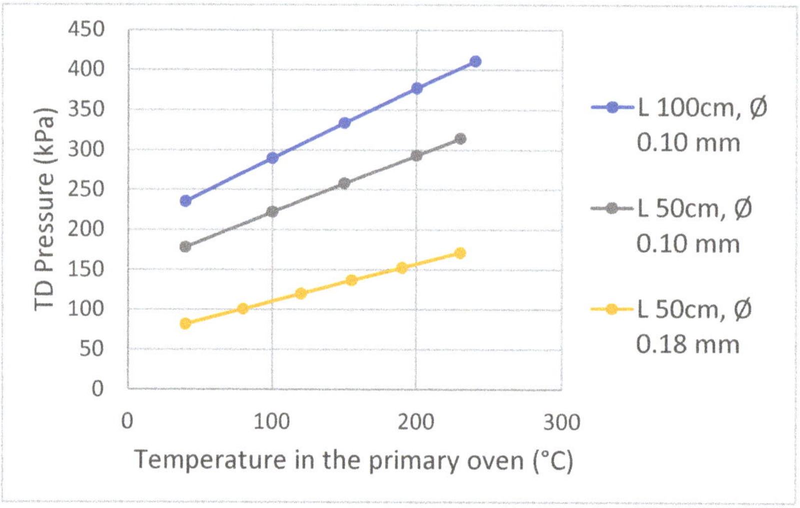

Figure 2. The influence of the secondary column dimension on the TD pressure.

Following the same logical process as in Part A, the performances of the three column configurations were compared for three materials in terms of the number of peaks observed: a trunk mat (multi-layer wood and felt), a PU foam, and a textile. The results are presented in Figure 3. Similarly to Figure 1, the three chosen thresholds share similar behaviours; therefore, only the threshold of 800 is represented here.

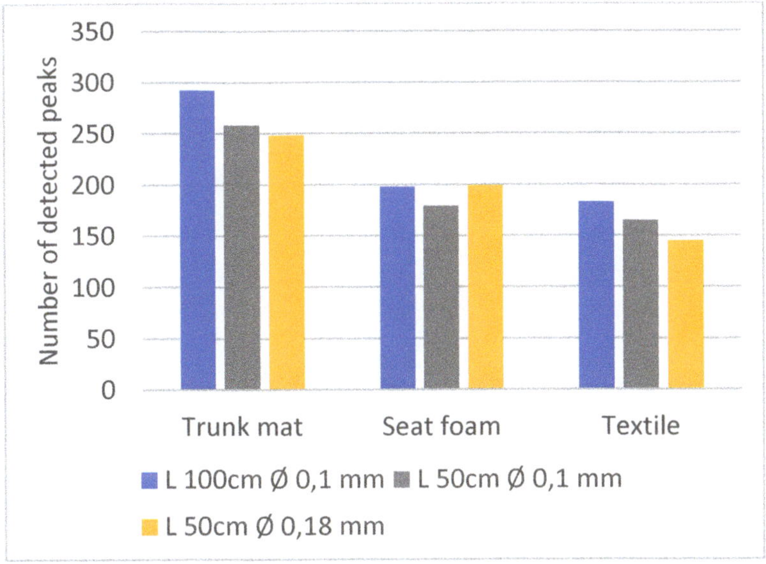

Figure 3. Number of potential odorants detected at similarity threshold of 800 depending on second column dimensions.

Of the three configurations, the 100 cm long and 0.1 mm wide one (the closest to the Part A protocol) permitted us to observe the greatest number of peaks, with 224 on average. This result is consistent with the fact that thin-film and small-diameter columns provide high efficiency at high gas velocity, thus limiting the loss of performance [36]. However, the 50 cm long and 0.18 mm wide column performed decently, with 197 peaks observed on average.

This loss of performance, amounting to 12% in terms of the average number of observed peaks in comparison to the best column, can be considered acceptable in regard to the benefit for the chromatographic system in the long term. Therefore, the shortest and widest secondary column was eventually selected. This completed the 1D-GC × GC-TOFMS hardware hyphenation.

In the end, the protocol developed in this part was considered to have been validated for the study of car material emissions. Its application to another sample of polypropylene allowed for the observation of 858 peaks, including hydrocarbons, out of which 144 were considered potentially odorous molecules, and 113 were reported to be odorous (Figure 4a). With such a large number of odorants, it may prove difficult to explain global odour with only a few of them. Moreover, these odorous molecules were widespread on the chromatogram, and it was not possible to restrict them to a few families or a small contour plot. In this chromatogram, the apparent tailing in the ^{2}D dimension is due to imperfectly modulated concentrated hydrocarbons and the choice of high contrast for the colour plot. This had a very limited impact on peak detection and identification.

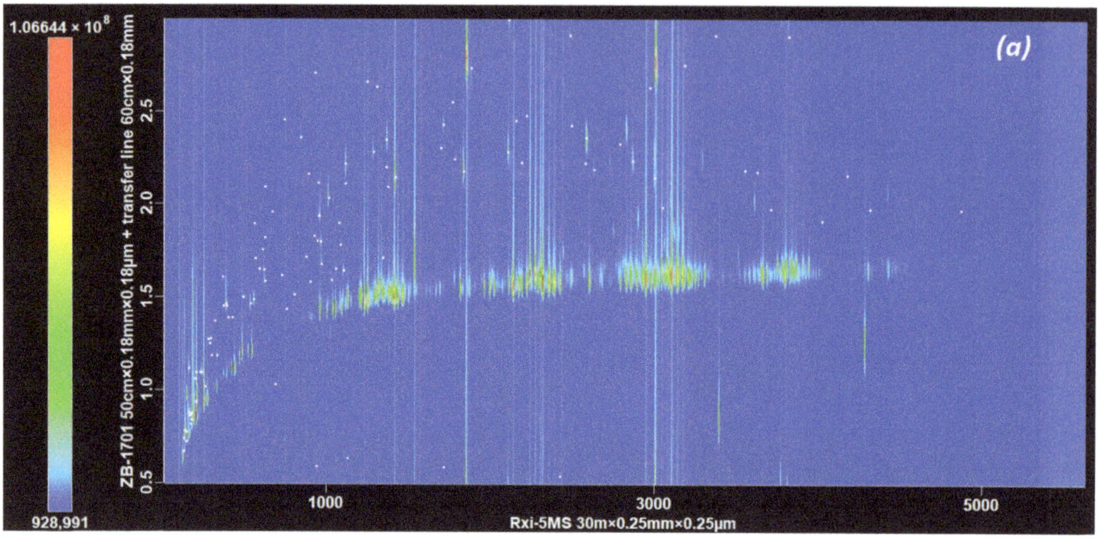

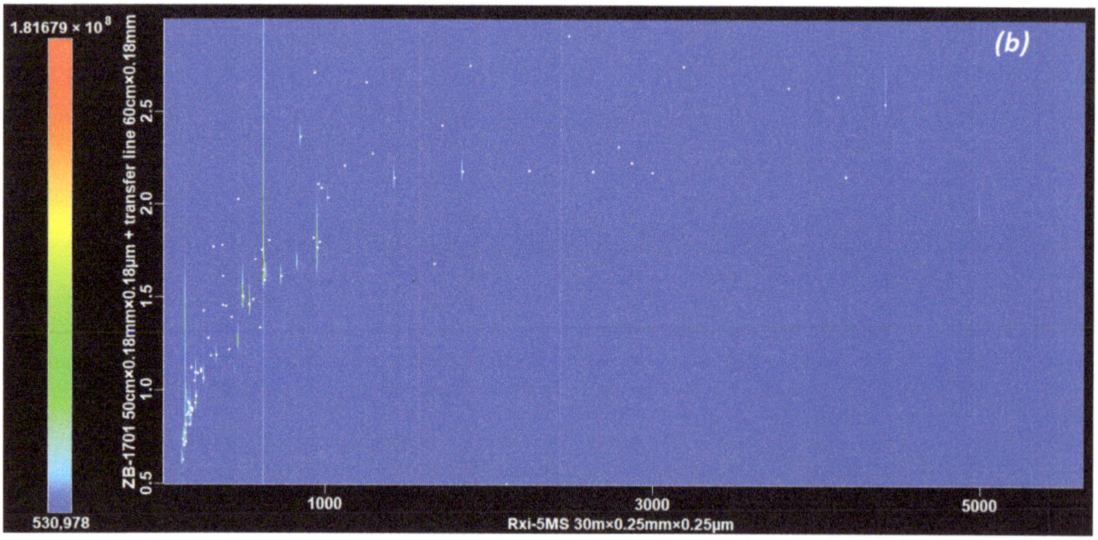

Figure 4. *Cont.*

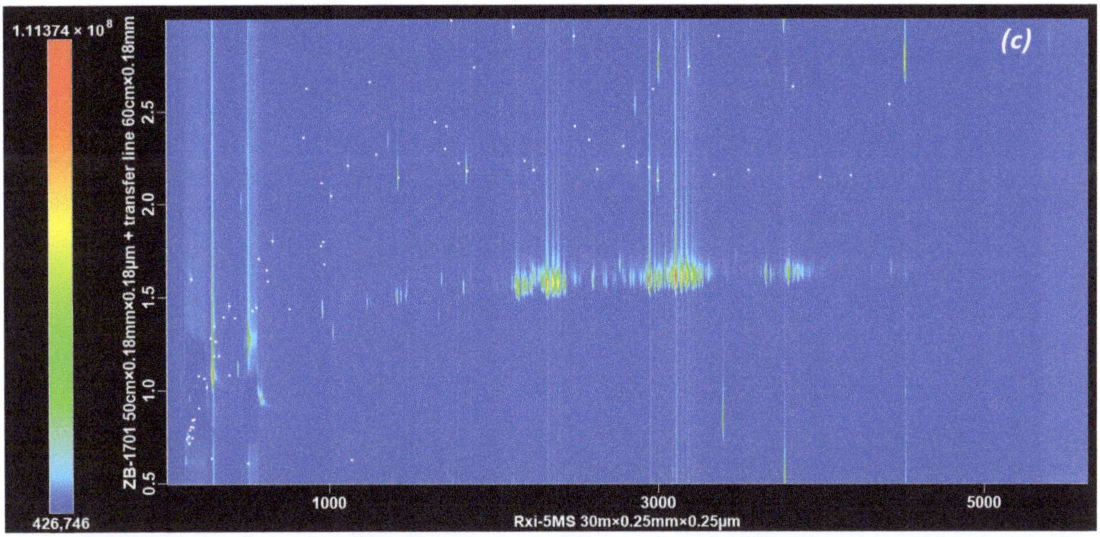

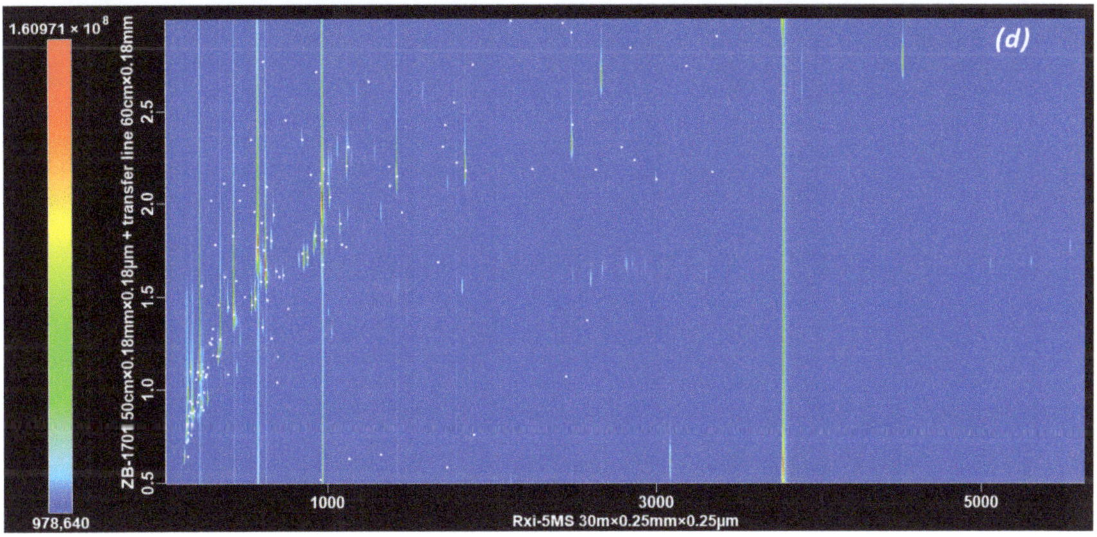

Figure 4. Chromatograms of 4 materials with the validated protocol: (**a**) PP, (**b**) ABS, (**c**) dashboard, and (**d**) décor. x axis: ^{1}D retention time (s). y axis: ^{2}D retention time (s). Each white square represents one of the identified odorant (75 in dashboard and ABS, 113 in PP, and 143 in dashboard).

The same protocol was also applied to three other materials (one ABS, one decor, and one dashboard, chromatograms in Figure 4b–d) for the search of odorous molecules. This was achieved by comparing the potential odorants retained to chemical databases, primarily Pubchem and The Good Scents Company. Including the aforementioned PP, between 75 and 147 odorous molecules were confirmed to be emitted by these four materials. The 20 most abundant molecules (based on the peak area on the chromatogram) for each material are presented in Table 2. As the exhaustive identification of odorants is very time-consuming, it was conducted on very limited samples, none of them being studied during Part A or Part B.

These last results are a preliminary proof of concept of the variety of molecules the method developed allow us to observe. During our experiments, we observed that the response factor between different molecules may vary more than an order of magnitude, which may alter their relative abundance. Moreover, peaks with an area above 1×10^9 may saturate the detector. Thus, without individual calibration curves, these areas must be considered as indicative. According to the areas measured during system suitability tests with a mixture containing 1 to 10 ng of 48 molecules, the individual compound amount was estimated ranging from 10 to 1000 ng.

Table 2. Lists of the 20 most abundant (based on peak area) odorous molecules detected in the emissions of (**a**) PP, (**b**) ABS, (**c**) dashboard, and (**d**) décor.

		(a) PP		
Name	**CAS Number**	**Area**	**^{1}D Retention Time (s)**	**^{2}D Retention Time (s)**
Tetrahydrofuran	109-99-9	5.31×10^8	169.989	0.884
Diphenyl ether	101-84-8	3.89×10^8	2577.336	2.362
Benzoic acid, methyl ester	93-58-3	3.04×10^8	1369.914	2.351
Nonanal	124-19-6	2.06×10^8	1419.912	2.145
Ethylbenzene	100-41-4	1.96×10^8	527.466	1.468
1-Butanol	71-36-3	1.94×10^8	194.9874	1.099
acetone	67-64-1	1.92×10^8	124.992	0.79
Benzene	71-43-2	1.83×10^8	194.9874	0.939
Naphthalene	91-20-3	1.71×10^8	1712.388	2.43
3-Penten-2-one, 4-methyl-	141-79-7	1.56×10^8	377.4756	1.471
1-Hexanol, 2-ethyl-	104-76-7	1.54×10^8	1117.428	2.218
p-Xylene	106-42-3	1.54×10^8	549.9648	1.49
1-Decanol	112-30-1	1.50×10^8	2104.866	2.283
Decanal	112-31-2	1.42×10^8	1834.884	2.179
Pyridine, 2,4,6-trimethyl-	108-75-8	1.33×10^8	954.936	1.942
Acetic acid	64-19-7	9.30×10^7	147.4908	1.119
1-Dodecanol	112-53-8	7.23×10^7	2864.814	2.237
Phenol	108-95-2	7.17×10^7	944.94	1.804
Benzaldehyde, 3,4-dimethyl-	5973-71-7	7.16×10^7	1882.38	2.691
Furan, 2,3-dihydro-	1191-99-7	6.91×10^7	147.4908	0.801
		(b) ABS		
Name	**CAS Number**	**Area**	**^{1}D Retention Time (s)**	**^{2}D Retention Time (s)**
Styrene	100-42-5	2.83×10^9	612.462	1.652
Phenol	108-95-2	2.29×10^9	947.442	1.768
2-Propenenitrile	107-13-1	1.89×10^9	132.4914	0.821
Benzene, chloro-	108-90-7	1.47×10^9	482.469	1.509
Ethylbenzene	100-41-4	9.96×10^8	524.9664	1.464
Methyl methacrylate	80-62-6	6.08×10^8	242.4846	1.069
Benzene, (1-methylethyl)-	98-82-8	4.32×10^8	719.952	1.614
Toluene	108-88-3	3.92×10^8	319.9794	1.193
Ethyl Acetate	141-78-6	3.68×10^8	162.4896	0.871
1-Butanol	71-36-3	2.84×10^8	194.9874	1.101
Propanenitrile	107-12-0	2.50×10^8	147.4908	0.941
Benzaldehyde	100-52-7	2.47×10^8	837.444	2.366
Benzene	71-43-2	1.79×10^8	194.9874	0.936
Carbon disulfide	75-15-0	1.71×10^8	134.9916	0.731
2-butanone	78-93-3	1.65×10^8	154.9902	0.888
Pyrrole	109-97-7	1.44×10^8	302.4804	1.773
Methane, dichloro-	75-09-2	1.10×10^8	132.4914	0.776
Nonanal	124-19-6	8.88×10^7	1419.912	2.146
Formic acid	64-18-6	8.02×10^7	119.9922	0.741
Acetic acid	64-19-7	7.79×10^7	167.4894	1.125

Table 2. *Cont.*

(c) Dashboard				
Name	**CAS Number**	**Area**	**^{1}D Retention Time (s)**	**^{2}D Retention Time (s)**
Pyridine	110-86-1	2.01×10^8	284.982	1.351
Furfural	98-01-1	1.39×10^8	454.971	2.009
Acetic acid	64-19-7	1.21×10^8	149.9904	1.604
Nonanal	124-19-6	1.16×10^8	1419.912	2.146
Decanal	112-31-2	8.78×10^7	1834.884	2.182
Dodecanoic acid	143-07-7	8.49×10^7	3184.794	2.738
Octanoic acid	124-07-2	7.95×10^7	1744,89	2.991
Nonanoic acid	112-05-0	7.93×10^7	2117.364	2.956
Sulphur dioxide	7446-09-5	6.95×10^7	494.9682	0.61
Decanoic acid	334-48-5	6.89×10^7	2484.84	2.904
1-Hexanol, 2-ethyl-	104-76-7	5.61×10^7	1117.428	2.209
Formic acid	64-18-6	5.13×10^7	117.4926	0.744
Hexanal	66-25-1	4.76×10^7	379.9758	1.453
2-Tridecanone	593-08-8	4.52×10^7	2944.812	2.201
Furan, 2,5-dimethyl-	625-86-5	4.15×10^7	242.4846	1.021
1-Dodecanol	112-53-8	3.72×10^7	2867.316	2.228
2(5H)-Furanone	497-23-4	3.48×10^7	744.954	0.959
Benzaldehyde	100-52-7	3.04×10^7	839.946	2.36
Carbon disulfide	75-15-0	2.99×10^7	134.9916	0.72
Heptanal	111-71-7	2.80×10^7	647.46	1.806
(d) Decor				
Name	**CAS Number**	**Area**	**^{1}D retention Time (s)**	**^{2}D Retention Time (s)**
1-Methoxy-2-propyl acetate	108-65-6	6.14×10^9	562.464	1.772
Styrene	100-42-5	3.27×10^9	612.462	1.651
Acetic acid, butyl ester	123-86-4	2.81×10^9	412.4736	1.409
2-Propanol, 1-methoxy-	107-98-2	2.66×10^9	212.4864	1.08
Isobutyl acetate	110-19-0	2.43×10^9	332.4786	1.273
Ethyl Acetate	141-78-6	1.80×10^9	164.9892	0.879
Benzene, 1,2,4-trimethyl-	95-63-6	1.38×10^9	964.938	1.812
p-Xylene	106-42-3	1.19×10^9	549.9648	1.494
Triethylamine	121-44-8	1.16×10^9	217.4862	0.915
Toluene	108-88-3	1.08×10^9	319.9794	1.186
Acetic acid	64-19-7	1.08×10^9	149.9904	1.107
2-Propenenitrile	107-13-1	1.06×10^9	129.9918	0.826
o-Xylene	95-47-6	8.42×10^8	617.46	1.503
Ethylbenzene	100-41-4	8.21×10^8	527.466	1.461
Mesitylene	108-67-8	6.09×10^8	874.944	1.713
Nonanal	124-19-6	5.89×10^8	1419.912	2.149
1-Hexanol, 2-ethyl-	104-76-7	4.74×10^8	1117.428	2.206
α-Methylstyrene	98-83-9	3.93×10^8	924.942	1.827
Benzene, 1,2,3-trimethyl-	526-73-8	3.74×10^8	1074.93	1.892
Ethanol, 2-butoxy-	111-76-2	2.95×10^8	659.958	1.946

4. Conclusions

In this study, we described the successful hyphenation between a Markes TD100-xr thermodesorber and a LECO Pegasus BT4D GC × GC-TOFMS device. Their combination is a first, and proper communication between the two instruments was not trivial and required some tricks. Moreover, without optimisation, our instrumental configuration resulted in the subjection of the thermodesorber to high pressures, close to its recommended limit. The first experimental protocol achieved was adapted by shortening and enlarging the ^{2}D column in order to reduce the global pressure drop. According to our global objective of mapping the potential odorants observed in vehicle material emissions, a special criterion was established to rate the separation performance. It consisted of enumerating the number of non-aliphatic hydrocarbon peaks, associated with their quality, which was assessed using their similarity score in mass spectrometry. An acceptable balance had to be found between the performances in terms of the number of compounds detected and the sustainability of the system, with a pressure that had more than halved at the cost of 12% fewer observed peaks.

The two main parameters selected for the final protocol were the phase combination of the two columns (5%-diphenyl and 14%-cyanopropylphenyl) and the dimensions of the ^{2}D column (50 cm × 0.18 mm × 0.18 μm). This experimental approach allowed us to observe several hundreds of polar potentially odorous molecules while keeping the pressure at rest below 100 kPa.

These results lay the foundation for further experiments on the observed molecules, starting with broader screening and the construction of a comprehensive odorant database. As every step of this study involved the selection of representative vehicle materials, our protocol should be suitable for any material used in this field, maybe even for synthetic materials of other origins, but this prospect remains to be confirmed by considering a wider range of other materials, including recycled ones. For further study, we expect to confront the global observed odorants with the olfactory perception of humans.

Supplementary Materials: The following supporting information can be downloaded at https://www.mdpi.com/article/10.3390/separations11060162/s1, Figure S1: TD-GC × GC-MS chromatogram of 48 VOC standard; Figure S2: Chromatograms of PP and leather with different column configurations.

Author Contributions: Conceptualisation, D.T. and G.G.; investigation, R.K.; resources, D.B., G.C. and J.D.; writing—original draft, R.K.; writing—review and editing, R.K., D.T., J.V. and G.C.; supervision, G.G., D.B., D.T. and J.V. All authors have read and agreed to the published version of the manuscript.

Funding: This work was funded by the ANRT under the CIFRE agreement No. 2019/0544.

Data Availability Statement: Data are available on demand.

Conflicts of Interest: Author Romain Klein, Guy Colombet, Donatien Barreteau and Guillaume Gruntz were employed by the company Renault Group. The remaining authors declare that the research was conducted in the absence of any commercial or financial relationships that could be construed as a potential conflict of interest.

References

1. Brancher, M.; Griffiths, K.D.; Franco, D.; De Melo Lisboa, H. A review of odour impact criteria in selected countries around the world. *Chemosphere* **2017**, *168*, 1531–1570. [CrossRef] [PubMed]
2. Zhang, G.-S.; Li, T.-T.; Luo, M.; Liu, J.-F.; Liu, Z.-R.; Bai, Y.-H. Air pollution in the microenvironment of parked new cars. *J. Affect. Disord.* **2008**, *43*, 315–319. [CrossRef]
3. Faber, J.; Brodzik, K.; Gołda-Kopek, A.; Łomankiewicz, D. Benzene, toluene and xylenes levels in new and used vehicles of the same model. *J. Environ. Sci.* **2013**, *25*, 2324–2330. [CrossRef] [PubMed]
4. Brodzik, K.; Faber, J.; Łomankiewicz, D.; Gołda-Kopek, A. In-vehicle VOCs composition of unconditioned, newly produced cars. *J. Environ. Sci.* **2014**, *26*, 1052–1061. [CrossRef] [PubMed]
5. Yoshida, T.; Matsunaga, I. A case study on identification of airborne organic compounds and time courses of their concentrations in the cabin of a new car for private use. *Environ. Int.* **2006**, *32*, 58–79. [CrossRef] [PubMed]
6. Yoshida, T.; Matsunaga, I.; Tomioka, K.; Kumagai, S. Interior Air Pollution in Automotive Cabins by Volatile Organic Compounds Diffusing from Interior Materials: I. Survey of 101 Types of Japanese Domestically Produced Cars for Private Use. *Indoor Built Environ.* **2006**, *15*, 425–444. [CrossRef]
7. Yoshida, T.; Matsunaga, I.; Tomioka, K.; Kumagai, S. Interior Air Pollution in Automotive Cabins by Volatile Organic Compounds Diffusing from Interior Materials: II. Influence of Manufacturer, Specifications and Usage Status on Air Pollution, and Estimation of Air Pollution Levels in Initial Phases of Delivery as a New Car. *Indoor Built Environ.* **2006**, *15*, 445–462. [CrossRef]
8. Janicka, A.B.; Reksa, M.; Sobianowska-Turek, A. The impact of car vehicle class on volatile organic compounds (VOC's) concentration in microatmosphere of car cabin. *J. KONES* **2010**, *17*, 207–212.
9. Chen, X.; Feng, L.; Luo, H.; Cheng, H. Analyses on influencing factors of airborne VOCS pollution in taxi cabins. *Environ. Sci. Pollut. Res.* **2014**, *21*, 12868–12882. [CrossRef]
10. Xu, B.; Chen, X.; Xiong, J. Air quality inside motor vehicles' cabins: A review. *Indoor Built Environ.* **2016**, *27*, 452–465. [CrossRef]

11. Buchecker, F.; Loos, H.M.; Buettner, A. Investigations on the impact of hardening on the odour of an aqueous cavity preservation for automotive applications using sensory and instrumental analysis. *Talanta Open* **2022**, *5*, 100095. [CrossRef]
12. VDA 276-3 (05/2022), VDA Webshop. Available online: https://webshop.vda.de/VDA/vda-276-3-052022 (accessed on 6 December 2022).
13. *ISO 12219-1:2021*; Interior Air of Road Vehicles—Part 1: Whole Vehicle Test Chamber—Specification and Method for the Determination of Volatile Organic Compounds in Cabin Interiors. ISO: Geneva, Switzerland. Available online: https://www.iso.org/standard/72410.html (accessed on 15 May 2022).
14. Buchecker, F.; Baum, A.; Loos, H.M.; Buettner, A. Follow your nose—Traveling the world of odorants in new cars. *Indoor Air* **2022**, *32*, e13014. [CrossRef]
15. Boyle, J.A.; Djordjevic, J.; Olsson, M.J.; Lundström, J.N.; Jones-Gotman, M. The Human Brain Distinguishes between Single Odorants and Binary Mixtures. *Cereb. Cortex* **2008**, *19*, 66–71. [CrossRef]
16. Laing, D.G.; Eddy, A.; Best, D.J. Perceptual characteristics of binary, trinary, and quaternary odor mixtures consisting of unpleasant constituents. *Physiol. Behav.* **1994**, *56*, 81–93. [CrossRef] [PubMed]
17. Jaubert, J.-N.; Gordon, G.; Dore, J.-C. Une organisation du champ des odeurs. I: Recherche de critères objectifs. *Parfums Cosmét. Arômes* **1987**, *77*, 53–56.
18. Jaubert, J.-N.; Gordon, G.; Dore, J.-C. Une organisation du champ des odeurs. II: Modèle descriptif de l'organisation de l'espace odorant. *Parfums Cosmét. Arômes* **1987**, 71–82.
19. Verriele, M.; Plaisance, H.; Vandenbilcke, V.; Locoge, N.; Jaubert, J.N.; Meunier, G. Odor evaluation and discrimination of car cabin and its components: Application of the "field of odors" approach in a sensory descriptive analysis. *J. Sens. Stud.* **2012**, *27*, 102–110. [CrossRef]
20. Agapakis, C.M.; Tolaas, S. Smelling in multiple dimensions. *Curr. Opin. Chem. Biol.* **2012**, *16*, 569–575. [CrossRef] [PubMed]
21. Kaeppler, K.; Mueller, F. Odor Classification: A Review of Factors Influencing Perception-Based Odor Arrangements. *Chem. Senses* **2013**, *38*, 189–209. [CrossRef]
22. Brattoli, M.; Cisternino, E.; Dambruoso, P.R.; De Gennaro, G.; Giungato, P.; Mazzone, A.; Palmisani, J.; Tutino, M. Gas Chromatography Analysis with Olfactometric Detection (GC-O) as a Useful Methodology for Chemical Characterization of Odorous Compounds. *Sensors* **2013**, *13*, 16759–16800. [CrossRef]
23. Delahunty, C.M.; Eyres, G.; Dufour, J.-P. Gas chromatography-olfactometry. *J. Sep. Sci.* **2006**, *29*, 2107–2125. [CrossRef] [PubMed]
24. Giungato, P.; Di Gilio, A.; Palmisani, J.; Marzocca, A.; Mazzone, A.; Brattoli, M.; Giua, R.; de Gennaro, G. Synergistic approaches for odor active compounds monitoring and identification: State of the art, integration, limits and potentialities of analytical and sensorial techniques. *TrAC Trends Anal. Chem.* **2018**, *107*, 116–129. [CrossRef]
25. Mondello, L.; Tranchida, P.Q.; Dugo, P.; Dugo, G. Comprehensive two-dimensional gas chromatography-mass spectrometry: A review. *Mass Spectrom. Rev.* **2008**, *27*, 101–124. [CrossRef] [PubMed]
26. Murray, J.A. Qualitative and quantitative approaches in comprehensive two-dimensional gas chromatography. *J. Chromatogr. A* **2012**, *1261*, 58–68. [CrossRef]
27. Lee, A.L.; Bartle, K.D.; Lewis, A.C. A Model of Peak Amplitude Enhancement in Orthogonal Two-Dimensional Gas Chromatography. *Anal. Chem.* **2001**, *73*, 1330–1335. [CrossRef]
28. Czerny, M.; Brueckner, R.; Kirchhoff, E.; Schmitt, R.; Buettner, A. The Influence of Molecular Structure on Odor Qualities and Odor Detection Thresholds of Volatile Alkylated Phenols. *Chem. Senses* **2011**, *36*, 539–553. [CrossRef]
29. Chastrette, M. Data Management in Olfaction Studies. *SAR QSAR Environ. Res.* **1998**, *8*, 157–181. [CrossRef] [PubMed]
30. Blomberg, J.; Schoenmakers, P.J.; Beens, J.; Tijssen, R. Compehensive two-dimensional gas chromatography (GC × GC) and its applicability to the characterization of complex (petrochemical) mixtures. *J. High Resolut. Chromatogr.* **1997**, *20*, 539–544. [CrossRef]
31. Ochiai, N.; Sasamoto, K. Selectable one-dimensional or two-dimensional gas chromatography–olfactometry/mass spectrometry with preparative fraction collection for analysis of ultra-trace amounts of odor compounds. *J. Chromatogr. A* **2010**, *1218*, 3180–3185. [CrossRef]
32. *ISO 12219-3:2012*; Interior Air of Road Vehicles—Part 3: Screening Method for the Determination of the Emissions of Volatile Organic Compounds from Vehicle Interior Parts and Materials—Micro-Scale Chamber Method. ISO: Geneva, Switzerland. Available online: http://www.iso.org/cms/render/live/en/sites/isoorg/contents/data/standard/05/48/54866.html (accessed on 30 August 2019).
33. Jores, C.D.S.; Klein, R.; Legendre, A.; Dugay, J.; Thiébaut, D.; Vial, J. Preparation of Thermodesorption Tube Standards: Comparison of Usual Methods Using Accuracy Profile Evaluation. *Separations* **2022**, *9*, 226. [CrossRef]
34. Wilson, R.B.; Siegler, W.C.; Hoggard, J.C.; Fitz, B.D.; Nadeau, J.S.; Synovec, R.E. Achieving high peak capacity production for gas chromatography and comprehensive two-dimensional gas chromatography by minimizing off-column peak broadening. *J. Chromatogr. A* **2011**, *1218*, 3130–3139. [CrossRef]
35. Shellie, R.; Marriott, P.; Morrison, P.; Mondello, L. Effects of pressure drop on absolute retention matching in comprehensive two-dimensional gas chromatography. *J. Sep. Sci.* **2004**, *27*, 503–512. [CrossRef]
36. Mommers, J.; van der Wal, S. Column Selection and Optimization for Comprehensive Two-Dimensional Gas Chromatography: A Review. *Crit. Rev. Anal. Chem.* **2021**, *51*, 183–202. [CrossRef]

separations

MDPI

Article

Combination of Natural Deep Eutectic Solvents and Nano-Liquid Chromatography towards White Analytical Chemistry: A Practical Application

Álvaro Santana-Mayor [1,2], Giovanni D'Orazio [2,*], Salvatore Fanali [3], Miguel Ángel Rodríguez-Delgado [1] and Bárbara Socas-Rodríguez [1,*]

[1] Departamento de Química, Área de Química Analítica, Facultad de Ciencias, Universidad de La Laguna (ULL), Avenida Astrofísico Francisco Sánchez s/n, 38206 San Cristóbal de La Laguna, Tenerife, Spain; asantanm@ull.edu.es (Á.S.-M.); mrguez@ull.edu.es (M.Á.R.-D.)
[2] Istituto per i Sistemi Biologici (ISB), CNR—Consiglio Nazionale delle Ricerche, Montelibretti, 00185 Rome, Italy
[3] School in Nanoscience and Advanced Technologies, University of Verona, 37129 Verona, Italy; salvatore.fanali@gmail.com
* Correspondence: giovanni.dorazio@cnr.it (G.D.); bsocasro@ull.edu.es (B.S.-R.); Tel.: +34-922-318-036 (B.S.-R.)

Abstract: In this work, a green and practical analytical method based on natural deep eutectic solvents (NADES) as extraction agents and nano-liquid chromatography as a separation technique was developed. To demonstrate the applicability of the methodology, alkylphenols and bisphenol A were evaluated as model compounds in olive and sunflower oils as model fatty samples by liquid–liquid microextraction. With this aim, several NADES based on mixtures of choline chloride with glycerol, lactic, ascorbic, and citric acids or glycerol with amino acids were evaluated as potential extraction solvents. In addition, to select the most suitable stationary phase for the separation of this group of contaminants, some stationary phases were tested, including Pinnacle II phenyl, Cogent Bidentate C_{18}™, and XBridge® C_{18}. The last one provided the best performance with an analysis time of 11 min. To solve the problem of the compatibility of hydrophilic NADES with chromatographic systems without harming the solubility of analytes, different aqueous organic mixtures were tested. Methanol/water mixtures were the most suitable as an injection solvent. Finally, following the White Analytical Chemistry principles, different tools were used to evaluate the greenness, the practicality, and applicability of the method based on the Analytical Eco-Scale, the Analytical GREEnness metric approach, and the Blue Applicability Grade Index.

Keywords: sustainability; white analytical chemistry; green solvent; nano-liquid chromatography; green chemistry; miniaturization; natural compound; oil; phenol

1. Introduction

The fulfilment of the Sustainable Development Goals has become a hot topic [1]. One of the areas which can have a great influence on the achievement of sustainability is the chemistry sector. Sustainability in chemistry serves as a foundational principle, aiming to integrate ecological, economic, and social dimensions [2–4]. This involves the responsible use of resources, minimization of the environmental impact, and the development of chemical products and processes that contribute positively to both human well-being and the health of the planet. It sets the stage for subsequent disciplines to align with global efforts towards a more sustainable future [1,3].

Within the scope of analytical chemistry, green analytical chemistry emerges as a pivotal concept [5]. This approach focuses on minimizing the environmental footprint of analytical techniques. Key strategies include the reduction or elimination of hazardous reagents, the optimization of energy consumption, and the development of rapid and efficient analytical methods while optimizing resource utilization. Green analytical chemistry emphasizes the selection of environmentally friendly solvents, the reduction of waste generation, and the incorporation of innovative technologies to achieve more efficient and eco-friendly analyses [5]. By adopting these green principles, analytical chemists contribute to the overall sustainability of chemical analysis without compromising the quality of the results. The combination of green principles with analytical methodologies functionality characterizes a novel approach called white analytical chemistry [2]. This involves well-balanced methods that adopt cleaner sample preparation methods, which reduce solvent usage and waste generation, and the use of analytical instrumentation that enables the detection and quantification of target compounds at lower concentrations, minimizing the amount of sample and reagents required while retaining their functionality and usefulness [2].

In this regard, nano-liquid chromatography (nano-LC) shows many advantages as a separation technique. The use of capillary columns with internal diameters typically lower than 100 μm added to flow rates in the nanoliter per minute range results in an enhanced sensitivity and chromatographic resolution. In addition, the reduced flow rates, even when compared to ultra-high performance liquid chromatography, allow for solvent consumption and waste generation to diminish substantially [6–8]. The low flow rate is also an improvement, since it allows for good compatibility with mass spectrometry systems in which the effluent is fully transferred to the detector [9,10]. These features make nano-LC systems suitable for the analysis of small amounts of sample and/or compounds found at trace levels in a shorter analysis time than in conventional systems. Despite packaging materials for nano-LC capillary columns being often expensive, the small amounts required make it an advantage in terms of cost/durability [6–8]. Based on that, nano-LC is considered a more environmentally friendly technique.

The great popularity of deep eutectic solvents (DES) is due to, among other things, their intrinsic properties and versatility, making them promising candidates for a more environmentally friendly alternative to conventional solvents in many fields of application, including extraction and separation processes, electrochemistry, synthesis, biomass treatment, biocatalysis, metal processing, drug delivery, CO_2 capture, food industry, and energy storage, among others [11–19]. Among these characteristics is its good biodegradability, especially in the solvents formed only by compounds found in nature. These DES are known as natural DES (NADES). This fact makes them more environmentally friendly than many conventional solvents. Likewise, the possible use of compounds that come from renewable sources for the preparation of NADES is aligned with the principles of sustainable chemistry, reducing their environmental impact. Compared to some traditional solvents, NADES generally exhibit lower toxicity, minimizing potential health risks for both researchers and end-users [18,19]. In the field of analytical chemistry processes, other characteristics stand out, including a broad range of solubility, tunable properties through the adjustment of the components and their molar ratios, offering a customizable platform for specific applications, and great thermal stability. Thus, it has been widely demonstrated in several publications that they possess great potential as extraction agents [18–21]. Nevertheless, its high viscosity, although effective in avoiding processes such as oxidation or degradation of certain compounds, hinders its fluidic properties, handling, and mass transfer between sample and solvent. In addition, its compatibility with analysis systems represents an analytical challenge [19,20]. Therefore, studying the different ways of combining NADES with chromatographic systems has become a necessity. It suggests, for instance, the evaluation of new combinations of NADES components to reduce the viscosity and improve their properties or improve their volatility.

Here, a novel analytical methodology is proposed based on the use of hydrophilic NADES as extraction solvents for a liquid–liquid microextraction (LLME) and nano-LC–UV-Vis as determination technique. As hydrophobic NADES are ideal for the extraction of organic compounds from aqueous samples, hydrophilic ones are very promising, among others, for the analysis of fatty samples. In this regard, a wide group of alkylphenols (phenol, 2,4-dimethylphenol (2,4-DMP), 2,3,6-trimethylphenol (2,3,6-TMP), 4-tert-butylphenol (4-TBP), 4-sec-butylphenol (4-SBP), 4-tert-amylphenol (4-TAP), 4-n-hexylphenol (4-HexP), 4-tert-octylphenol (4-TOP), 4-n-heptylphenol (4-HepP), 4-n-octylphenol (4-OP), and 4-n-nonylphenol (4-NP)) and bisphenol A (BPA) were used as representative analytes, whereas olive and sunflower oils were selected as model matrices representing oily samples. Finally, different tools were applied to evaluate the greenness and blueness of the method.

2. Materials and Methods

2.1. Chemicals and Materials

Standards and solvents were used as received (without any additional purification process). The analytical standards of phenol (CAS 108-95-2), BPA (CAS 80-05-7), 2,4-DMP (CAS 105-67-9), 2,3,6-TMP (2416-94-6), 4-TAP (CAS 80-46-6), 4-OP (CAS 1806-26-4), 4-TOP (CAS 140-66-9), 4-NP (CAS 104-40-5) with a purity higher than 96.3% were purchased from Dr. Ehrenstorfer GmbH (Augsburg, Germany), while 4-TBP (CAS 98-54-4), 4-HexP (CAS 2446-69-7), and 4-HepP (CAS 1987-50-4) with a purity higher than 99.9% were supplied by Sigma-Aldrich Chemie (Madrid, Spain) and 4-SBP (CAS 99-71-8) with 98.0% of purity was acquired from Combi-Blocks, Inc. (San Diego, CA, USA). Stock solutions of studied compounds were prepared in methanol at 1000 mg/L and stored at −18 °C in darkness. Working solutions were prepared through appropriate dilutions.

Water, methanol, acetonitrile (can) for HPLC, and acetone (HiPerSolv CHROMANO-RM®) were from VWR International, Part of Avantor (Milan, Italy). Choline chloride (≥98.0%), citric acid (99.0%), glycerol (≥99.5%), and DL-lactic acid (≥85.5%) were from Sigma-Aldrich (Milan, Italy), whereas ascorbic acid (≥99.7%) was from Carlo Erba Reagents (Milan, Italy). L-Alanine (≥99.0%), L-histidine (≥99.0%), and L-lysine (≥97.0%) were from Fluka (Milan, Italy), while L-proline (99.0%), benzene, ethylbenzene, and propylbenzene were from Merck (Darmstadt, Germany).

Capillary columns were prepared by utilizing polyimide-coated fused silica capillaries purchased from Polymicro Technologies (West Yorkshire, UK), featuring OD of 375 μm and an ID of 100 μm. Different stationary phases were investigated, including bulk-acquired Pinnacle II phenyl (3 μm, 110 Å pore size, 6% carbon load) from Restek (Bellefonte, PA, USA), Cogent Bidentate C_{18}™ (4 μm, 100 Å pore size, 18–19% carbon load) obtained from MicroSolv (Leland, NC, USA), and XBridge® C_{18} (3.5 μm, 130 Å pore size, 18% carbon load) sourced by unpacking a guard cartridge from Waters (Milford, MA, USA).

2.2. Sample Selection

In this work, commercially available olive and sunflower oils were purchased in local supermarkets in Rome, Italy, with the aim of using them as model samples of fatty foods.

2.3. Apparatus and Software

An FS 100b Decon ultrasonic bath (Hove, UK) was used to sonicate mobile phases, analyte solutions, and stationary phase slurry during the packing procedure. An optical transmission microscope, Stereozoom 4 Microscope with an illuminator device (Bausch & Lomb, Rochester, NY, USA) was used to check the status of the stationary phase into the column in the packing step. Finally, a Perkin Elmer HPLC Series 10 pump (Palo Alto, CA, USA) was employed for moving the stationary phase slurry into the fused silica capillary.

A nano-/capillary-LC pump, Ultimate™ Capillary HPLC unit from LC Packings Dionex (Amsterdam, The Netherlands), equipped with a manual sample injector and an internal flow splitting unit, was employed for nano-LC experiments. The injector device was a low dispersion six-port valve from VICI VALCO Instrumentations (Houston, TX, USA) equipped with a 15 μL external loop. To reduce dead volumes, thus minimizing the chromatographic band broadening effect, the capillary columns were directly connected to the injector valve.

The target compounds were detected by a laboratory-made on-column detection cell at 214 nm. The temperature of the chromatographic system was controlled by continuous room conditioning in the range of 18–22 °C. The pump and UV detector were controlled by Chromeleon™ Chromatography Management System Software (Version 6.6, LC Packings).

The separation process was carried out using a step gradient mode [22], regulated through the injection valve. The external loop was employed for both, loading analytes/samples and introducing the mobile phase. The injection volume was adjusted based on the injection time at the column workflow. In this setup, ACN, as the pump solvent, was exclusively used to thrust forward the injection plug into the column. The waste solvent was recycled into the pump reservoir [8]. Under the final chromatographic conditions, the step gradient mode was achieved by plug injection of aqueous/organic ratios with increased elution strength. The applied gradient is reported in Table 1. At the end of the day, the column was flushed with ACN. Forty nanoliters was the injection volume applied to all studied columns.

Table 1. Step gradients developed for the different stationary phases.

Stationary Phase	Column Conditioning	Step gradient (Time; Composition)
Pinnacle II phenyl	50/50, ACN/H_2O, *v*/*v*	8.1 min; 70/30, ACN/H_2O, *v*/*v*
Cogent Bidentate C_{18}™	50/50, ACN/H_2O, *v*/*v*	9.0 min; 80/20, ACN/H_2O, *v*/*v*
XBridge® C_{18}	70/30, ACN/H_2O, *v*/*v*	5.5 min; 90/10, ACN/H_2O, *v*/*v*

Injection volume: 40 nL; flow rate: 240 nL/min.

2.4. Preparation of Packed Capillaries

All capillary columns were prepared in our laboratory following a slurry packaging method [6] and packed with an effective and total length of 25.0 cm and 35.0 cm, respectively, as shown in Figure 1.

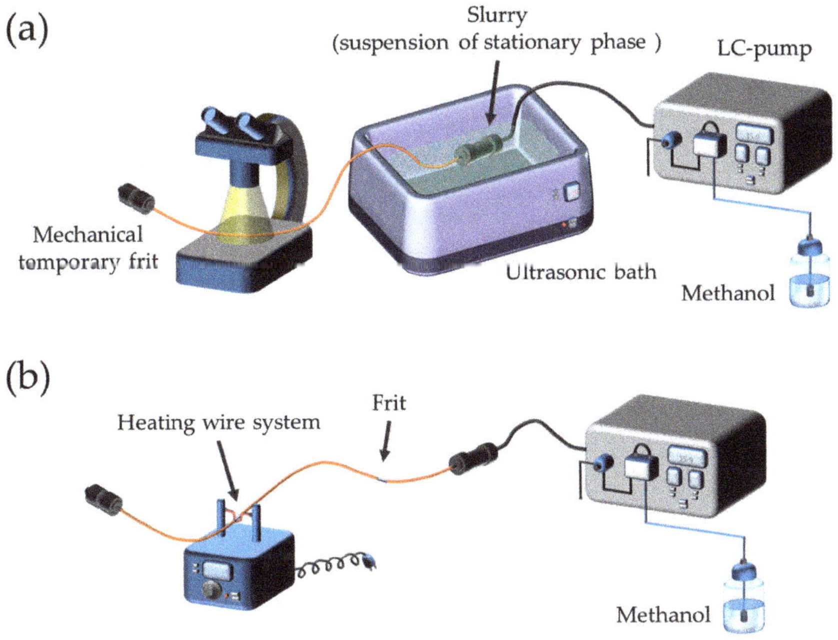

Figure 1. Scheme of (**a**) the system used for capillary packaging using the slurry method and (**b**) the preparation procedure for sintered silica frits.

Briefly, each end of 35–40 cm of fused silica capillary was joined both to a mechanical temporary frit (Valco, Houston, TX, USA) used to retain the packing material and to a 10 cm × 4.1 mm ID stainless steel HPLC pre-column (Valco) used as a reservoir for the slurry, respectively. Each slurry was prepared suspending about 50 mg of the selected stationary phase in 1 mL of acetone. After a brief sonic bath, the homogenous suspension obtained was transferred to the reservoir and pumped through the empty capillary by using an LC pump. At a maximum of 35 MPa (350 bar, 5000 psi), the capillary was packed for a length of about 30 cm. During this step, each capillary was immersed in the ultrasound bath in order to obtain a homogeneous stationary-phase bed structure. After a thorough washing of the phase with distilled water, the stationary phase was raised with 5 mM NaCl. A laboratory-made electrical heated wire was used to prepare inlet and outlet frits by simply heating a small capillary section at about 700 °C for 7–8 s. The temporary frit was removed, and the excess of stationary phase was eliminated in the back-flush mode.

The detection windows were prepared by removing the outer polyimide layer with a razor for a length of about 0.5 cm, close to the outlet frit (effective length, 26.6 cm). An optical microscope was used to monitor the whole procedure. Finally, each capillary was cut to the desired length and was equilibrated using a 50/50 (*v*/*v*) H_2O/ACN mixture.

A mixture of three alkylbenzenes, namely benzene, ethylbenzene, and propylbenzene (0.1% (*v*/*v*) each in water), was used as a test sample and periodically injected to evaluate the chromatographic performance of each column. Injected at 60 nL, the test mixture was separated in isocratic mode by using a mobile phase consisting of an ACN/water (80/20, *v*/*v*) mixture.

A comparative study of the main chromatographic parameters was the tool to assess the robustness of the system and the goodness of the prepared column.

2.5. NADES Preparation

The preparation of NADESs involved the magnetically assisted agitation of the components at specific molar ratios, according to previous publications, with a few modifications [7,23–25], as shown in Table 2. Glass tubes were then subjected to heating at 80 °C for 30 min, resulting in a homogeneous and clear liquid mixture. The resulting solvent was stored in a vacuum desiccator to prevent moisture until ready for use.

Table 2. Components for the synthesis of the different NADES used.

Component 1	Component 2	Component 3	Molar Ratio	Reference
Choline chloride	Glycerol	-	1:2	[23]
Choline chloride	Lactic acid	-	1:2	[24]
Choline chloride	Ascorbic acid	H_2O	2:1:4	[7]
Choline chloride	Citric acid	H_2O	2:1:4	[24]
Glycerol	L-Alanine	-	3:1	[25]
Glycerol	L-Histidine	-	3:1	[25]
Glycerol	L-Proline	-	3:1	[25]
Glycerol	L-Lysine	-	4.5:1	[25]

2.6. Microextraction Tests

One hundred microliters of NADES were rapidly added to 1 mL of oil sample. Then, the solvent was dispersed using one minute of vortex agitation. Later, the sample was centrifuged for one minute. An aliquot of the NADES-enriched phase placed at the bottom of the centrifuged tube (20 μL) was collected and diluted four-fold with a methanol/water mixture (50/50, *v*/*v*) to a final volume of 80 μL. The diluted sample extract was finally blended prior to its injection into the nano-LC system (Figure 2).

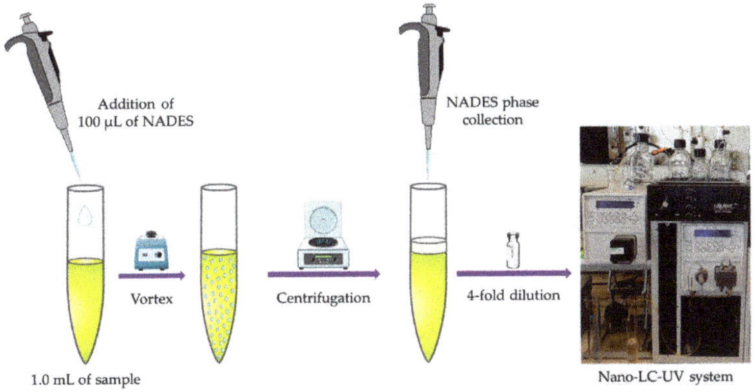

Figure 2. Schematic illustration of the NADES-LLME-nano-LC-UV procedure.

2.7. Sustainability Assessment Tools

For the evaluation of the environmental friendliness of the method, different tools were used. It included the Analytical Eco-Scale [26], the Analytical GREEnness (AGREE) metric approach [27], and, finally, the more recent Blue Applicability Grade Index (BAGI) [28], introducing a new metric tool for assessing the practicability of a method.

3. Results and Discussion

A significant effort has been undertaken to steer analytical chemistry toward a more sustainable paradigm aligned with the 12 principles of sustainable chemistry proposed by Anastas and Warner [5]. Key criteria, such as "preventing waste, using safer solvents and reaction conditions, and designing safer chemicals and products" are effectively addressed through the miniaturization of analytical methodologies. Novel proposals, for instance, include the application of nano-LC systems combined with miniaturized extraction techniques and the utilization of NADES as alternative extraction solvents.

3.1. Optimization of the Nano-LC-UV Method

For the development of a methodology that allowed the determination of alkylphenols and bisphenols in the samples of interest, firstly it was necessary to optimize the chromatographic separation.

The alkylphenols exhibit characteristics of being non-charged compounds in a wide range of pH with pKas higher than 9.5 and a hydrophobic molecular structure characterized by partition coefficients (log P) in the range 1.46–5.76 [29]. This final consideration suggests that the nature of the molecular structure imposes restrictions on the selection of the stationary phase, particularly in relation to the solvent used to dilute the standard mixture. From the perspective of environmental sustainability, an approach based on a reverse-phase partitioning mechanism was adopted, choosing stationary phases designed to enhance interaction with the target compounds. Based on this consideration, three reverse-phase stationary phases, two C_{18}, and one phenyl moieties linked to silica particles, totally end capped, were studied. Different parameters (i.e., flow rate, mobile phase composition, injection solvent) were changed in order to reach the best chromatographic separation in terms of performance and analysis time.

The miniaturization of the separation system using capillary columns provides two significant chromatographic advantages. Firstly, it promotes a greater homogeneity of the packing bed, thus reducing the effect of chromatographic band broadening [30]. Additionally, it contributes to a lower chromatographic dilution [31]. Given that the downscaling effect requires an injection volume of a few nanoliters, the extremely low analytical sensitivity it entails is certainly not compatible with the development of a novel analytical methodology. This limitation can be overcome by increasing the injection volume and by working under overloading conditions. Although this is a chromatographically disadvantageous condition, it can be controlled and transformed into a valid online pre-concentration technique defined as on-column focusing. This approach allows for the concentration/focusing of the analyte at the column head without affecting the column's performance and chromatographic profile [32].

Generally, in reverse-phase chromatography, during the injection step, when non-polar compounds are dissolved in mixtures with a lower elution strength compared to the initial mobile phase, thinning of the injection plug occurs at the head of the column. Increasing the water content in the solvent dilution has been found to be an easy solution to pre-concentrate and focus the on-column analytes, maintaining a chromatographic profile characterized by a controlled band broadening effect [33]. On the contrary, the injection of organic solvent (ACN, methanol) was found to result in a worse chromatographic profile despite the small volume used in this case (60 nL). In addition, the maximum effect of injection plug compression was achieved using water as the dilution solvent [33]. In this context, due to the low solubility in the aqueous phase, especially for the more hydrophobic alkylphenols (such as 4-TOP, 4-HepP, 4-OP, 4-NP), an aqueous organic mixture was necessary to obtain an appropriate chromatographic profile.

Initial experiments focused on enhancing solubility by exploring different aqueous organic mixtures. The analytical standards were diluted in basic aqueous mixtures of alcohol and ACN. Equivalent injection volumes resulted in higher chromatographic peaks, indicating increased sensitivity, when using ACN/water and methanol/water mixtures. Additionally, it was observed that the use of methanol allowed for improved chromatographic separation profiles on all studied columns in terms of performance and repeatability, as it has also been observed in previous works [34,35].

Following the step gradient described in Table 1, Figure 3 shows a comparative study of the selectivity of the stationary phases studied that provided the best baseline separation, but under different conditions and analysis times.

For silica C_{18}-packed columns, the elution order was maintained, but significant differences in selectivity were observed. Although XBridge C_{18} and the Bidentate C_{18} exhibit a similar carbon load (18–19%), the latter showed notably higher retention for 4-TAP, 4-HexP, 4-HepP, 4-OP, 4-NP (11–17 min). However, by using the XBridge C_{18} column, analytes could be eluted in 7–11 min by a final elution phase with 90% ACN. It may be ascribed to the slightly reduced particle size of XBridge C_{18} compared to the Bidentate C_{18}, resulting in a larger available surface area and, consequently, enhanced retentive properties influencing peak width. Furthermore, the greater selectivity of the XBridge stationary phase, resolving 12 out of 13 peaks at baseline compared to Bidentate, which resolved 10 out of 13 peaks, may be attributed to the different pore sizes. A pore size of 130 Å for XBridge compared to 110 Å for the Bidentate C_{18} likely allowed for better accommodation of steric bulkier alkylphenols, improving interaction and expediting the partitioning mechanism.

Finally, despite the low carbon load, the Pinnacle II Phenyl phase exhibited exceptional selectivity for alkylphenols, facilitated by the presence of unsaturated moieties promoting π–π interactions with the phenyl moieties of the stationary phase. Due to the different partitioning mechanisms, a reversal in the elution order of 2,4-DMP and BPA was observed compared to C_{18}, particularly in the separation of 4-MP and BPA, which coeluted in the XBridge column.

To achieve the optimal compromise between on-column focusing effects and analytical solubility, under the most suitable chromatographic separation conditions achieved, the mixture of alkylphenols was injected with different ratios of water/methanol mixture. The 50/50 (v/v) water/methanol ratio allowed for good sensitivity for each studied column but with changing effects in terms of chromatographic efficiency. The greater the elution strength difference between the dilution solvent and the conditioning mobile phase, the smaller the observed band broadening effect. The high selectivity and carbon load of

XBridge C_{18} enabled elution with a higher organic phase content, simultaneously achieving a thin injection plug even with a high methanol content.

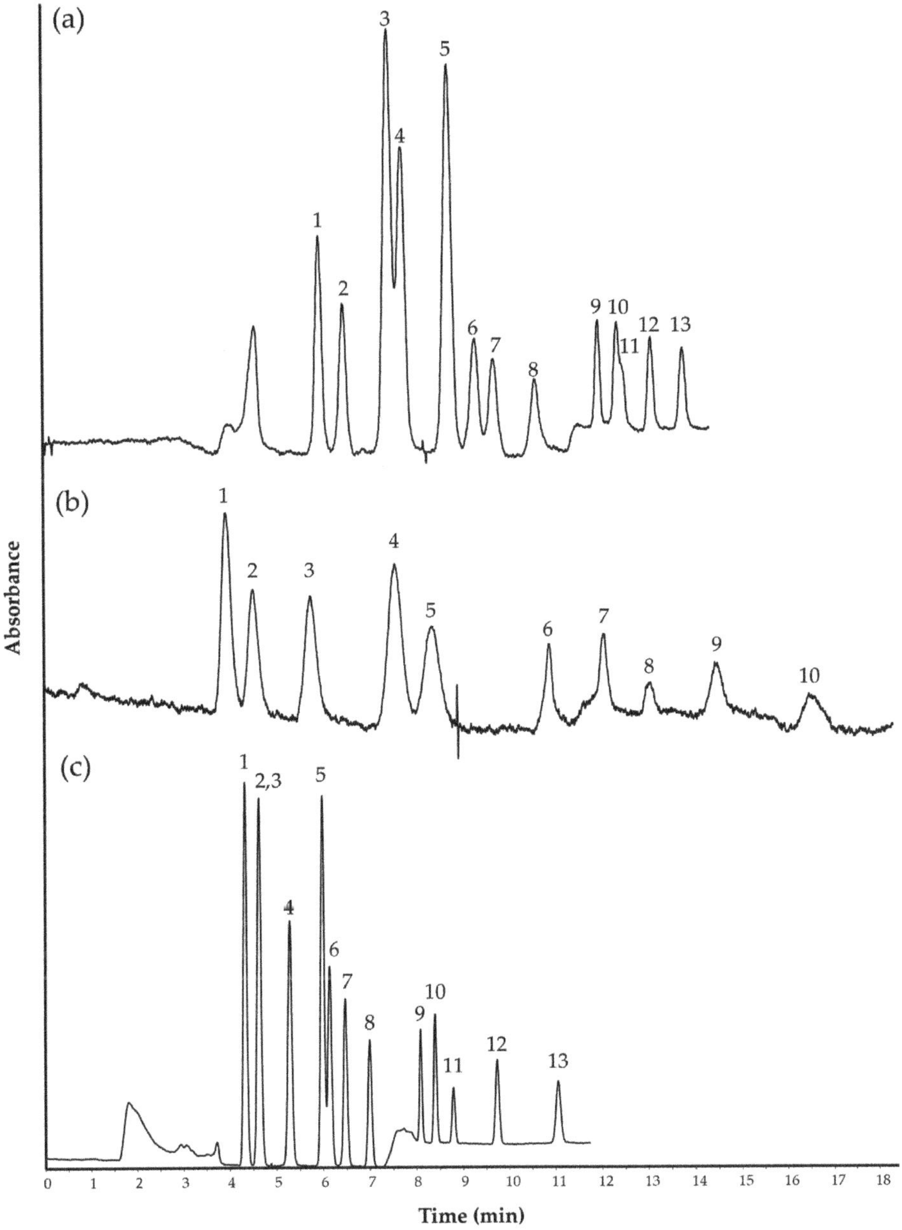

Figure 3. Selectivity of different stationary phases toward the studied compounds. Comparison of nano-LC-UV chromatograms (λ: 214 nm). Experimental conditions: analytes concentration at 16 µg/mL. (**a**) Pinnacle II Phenyl; (**b**) Cogent Bidentate C_{18}; (**c**) XBridge C_{18}. For gradient conditions, see Table 2. Peak identification: (1) Phenol, (2) 4-MP, (3) BPA, (4) 2,4-DMP, (5) 2,3,6-TMP, (6) 4-TBP, (7) 4-SBP, (8) 4-TAP, (9) 4-HexP, (10) 4-TOP, (11) 4-HepP, (12) 4-OP, (13) 4-NP.

Based on these results, as depicted in Figure 3, the XBridge C_{18} capillary column achieved the best outcomes in terms of selectivity, analysis time, and peak shape, and, thus, it was selected for further experiments.

3.2. NADES-LLME-Nano-LC-UV Method for the Analysis of Vegetable Oil Samples

Once the separation conditions were established, the method was applied to evaluate target analytes in two vegetable oils as model samples. Intending to develop a more sustainable methodology and considering the fact that the determination system complies with several points of the white analytical chemistry, the next step consisted of the performance of the extraction procedure. Based on that, nowadays, NADES are one of the alternatives that best fit the trends for the sake of more sustainable analytical chemistry approaches, including LLME methods [36]. Therefore, different NADES based on choline chloride with glycerol or organic acids (lactic, ascorbic, citric acids) and glycerol with different amino acids (L-alanine, L-histidine, L-proline, L-lysine) were prepared as described in Section 2.5.

Despite the many advantages that characterize NADES, sometimes these go against the development of the analytical methods. Among them, their viscosity, which makes the handling and direct injection of NADES extract difficult, and their low immiscibility with conventional organic solvents that are used both to dilute the extract and as components of the chromatographic system are two examples [19,20]. In this sense, in general terms, the NADES prepared in the present work are characterized by a low miscibility in solvents such as ACN. This implies that the dilution solvent must contain a high percentage of water when this type of NADES is used ($\geq$60%, *v*/*v*). On the one hand, this aspect benefits the chromatographic performance, since the elution strength of the injection solvent is lower. To achieve a good chromatographic separation, the injection solvent must have a lower elution strength than that of the mobile phase. Otherwise, not only is chromatographic efficiency and resolution lost, but it also does not favor the on-column focusing. On the other hand, the low solubility of target compounds in aqueous phases hindered the use of a high-water content mixture of solvents. Due to all of that, the optimization of the chromatographic system in combination with hydrophilic extracts and non-polar compounds represents a very difficult challenge.

In this case, and as previously mentioned, although alkylphenols are easily soluble in organic solvents, the NADES-based extracts are not miscible with those. For this reason, different dilution solvents were tested to find a compromise between all the factors that could be detrimental to the development of the method: chromatographic efficiency and resolution, sensitivity, solubility of the analytes, and compatibility of solvents, among others. Thus, it was found that the smaller alcohols, i.e., methanol, ethanol, and isopropanol, in proportions of 50/50, *v*/*v*, with water, are not miscible with NADES but also with solubilize alkylphenols. Furthermore, the chromatographic efficiency, resolution, and sensitivity were comparable to those of solvent standards, which greatly benefited chromatographic development. Despite this, and as expected, as the viscosity of the alcohol increases in the order isopropanol > ethanol > methanol, the viscosity of the injection solvent also increases; therefore, the chromatographic efficiency and resolution worsen. Therefore, the mixture formed by methanol and water in a 50/50 (*v*/*v*) proportion, was selected as the most suitable for further experiments.

After the optimization of the injection conditions, the different synthesized NADES were applied to the extraction of alkylphenols and BPA from olive and sunflower oils, as representative samples of oily nature, obtaining promising results. As it can be seen in Figure 4a, the chromatogram of a standard containing 25% (*v*/*v*) of NADES of choline chloride and glycerol in methanol/water (50/50, *v*/*v*) is represented. At the wavelength used, the NADES does not present the absorption band characteristic of the hydrophilic NADES, which usually elutes at the first part of the chromatogram [37,38]. However, in the oil extracts, shown in Figure 4b–d, a large band of more than a minute wide can be seen, and its intensity is much higher than that of the compounds of interest. As it can be observed, all tested NADES were able to extract most of the target compounds from the model matrix, which demonstrates their potential to extract polar compounds from fatty matrices, as has also been remarked in other works [39–41]. The most polar analytes included in this study (phenol and BPA) are very affected by the tail of the co-extracted matrix components. Nevertheless, in general terms, the application of the different evaluated NADES resulted in very good results. Indeed, their chromatographic performance is comparable with that of a solvent standard. Similar results were obtained for the other NADES described in Table 2, as well as for the sunflower matrix.

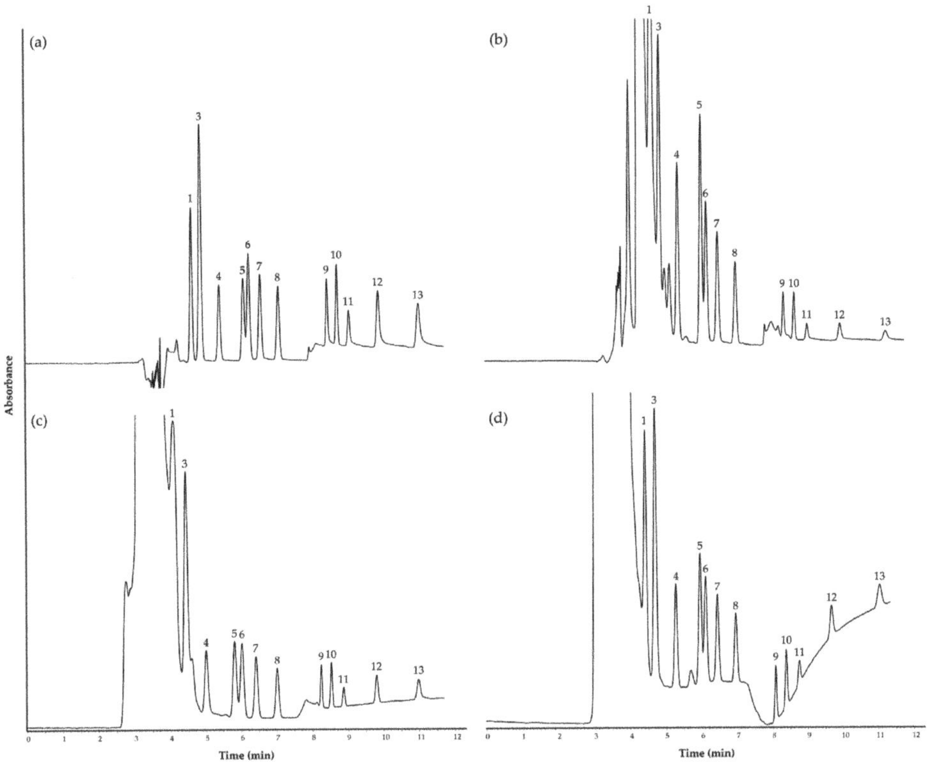

Figure 4. Nano-LC-UV separation of (**a**) a solvent standard containing 25% (v/v) of choline chloride:glycerol NADES in methanol:water (50/50, v/v) and olive oil after the microextraction step using a NADES of choline chloride and (**b**) glycerol, (**c**) lactic acid, and (**d**) ascorbic acid. For NADES composition and gradient conditions, see Tables 1 and 2. Peak identification: (1) Phenol, (3) BPA, (4) 2,4-DMP, (5) 2,3,6-TMP, (6) 4-TBP, (7) 4-SBP, (8) 4-TAP, (9) 4-HexP, (10) 4-TOP, (11) 4-HepP, (12) 4-OP, (13) 4-NP.

Additionally, it is worth mentioning that, although in this case there exists overlap between the matrix band and the most polar analytes that hinders their correct determination when a UV detector is used, the application of these kinds of conventional instruments increases the accessibility of the methodology to a greater number of laboratories. Apart from that, this approach could be improved for those laboratories that have a mass spectrometry detector, in which the resolution is based on the mass/charge ratio as well as on the fragmentation of the compounds, in the case of tandem mass spectrometry, as this approach could be of great importance in solving this type of problem.

3.3. Sustainability Assessment

Several tools have been suggested for evaluating the compliance of metric tools with the principles of the green analytical chemistry, including the National Environmental Methods Index (NEMI) [42], the Analytical Eco-Scale [19], the (Complementary) Green Analytical Procedure Index (GAPI) [43,44], the AGREE [27] and its version for sample preparation procedures [45], and, more recently, the BAGI [28] approaches. These methods have unique considerations and distinct frameworks for assessing their eco-friendly profile and, therefore, their results and interpretations can lead to debate and controversies about the environmental, useful, and practical validity of a methodology. To easily compare the different tools and evaluate the information of the results obtained, the Analytical Eco-Scale [26], AGREE [27], and BAGI [28] tools were compared.

The Analytical Eco-Scale [26] aims to evaluate the green nature of an analytical procedure in a "semiquantitative" way by assigning penalty points to those parameters of the analytical process that are detrimental to the ideal green analysis (represented by a score of 100). The AGREE method [27] evaluates the greenness of an analytical methodology based on the 12 principles of green analytical chemistry. The result is reflected in a pictogram divided into 12 parts whose color ranges from dark green (green principle) to red (non-green principle). In the center of the pictogram, the result appears reflected on a scale from 0 to 1, in which 1 represents the ideal analytical procedure. Finally, the BAGI [28] measures the practicality and applicability of the method. It was performed as a complementary metric tool to the green metrics considering the principles of white analytical chemistry. The response obtained is shown in a blue-colored asteroid pictogram divided into ten parts with the final score in the central section (on a scale of 25 to 100). In these terms, white analytical chemistry first appeared in 2021 as an alternative approach to green analytical chemistry [2]. This new approach, in addition to the green character of a method (greenness), proposes a red principle that reflects the validity of the analytical parameters of a method (redness) and a blue principle because of its practical attributes (blueness). Notwithstanding, the red part depends, to a large extent, on the analysis system used and the ranges evaluated relative to the figure of merits. Therefore, this aspect has not been included in the comparison. The results obtained are shown in Figure 5.

Based on the Analytical Eco-Scale tool (Figure 5a) [26], a score of 84 out of 100 was obtained for those procedures in which lactic- or citric-acid-based NADES were used. For the other NADES used as extraction solvents, 85 out of 100 points were obtained since no pictograms or signs of danger were presented. This final score represents an excellent green analysis. Nevertheless, some aspects must be explained in this sense. For instance, although the use of methanol is limited to 30 μL as a diluent of the final extract of the sample, the penalty points are the same for those volumes lower than 10 mL. Therefore, this score is overstated.

In the case of AGREE (Figure 5b) [27], a final score of 0.66 out of 1.00 was obtained for all cases. This result is in good agreement with some of the principles of green analytical chemistry. However, some aspects are difficult to comply with when attempting to analyze a large number of compounds in complex samples such as vegetable oils. Among these aspects is the unreal situation of analyzing this type of sample without prior treatment (point 1). This point is closely related to the type of analysis, which also makes direct analysis of the sample impossible (point 3). Furthermore, the need for systems that allow determinations in a selective and specific manner, such as that based on chromatographic systems, is detrimental to what is considered to be a green procedure (point 9). Likewise, as in the Analytical Eco-Scale, the use of solvents with hazard pictograms or that are not bio-based, even in very small quantities, severely impairs the evaluation. All these aspects present an unrealistic approach to analytical procedures; despite the greatest number of determinations possible being sought with families of compounds of different types in samples of a diverse nature, the only way to do this is with systems that use high energy and prior sample treatment.

(a)

Analytical Eco-Scale	
Reagents	**Penalty points**
Lactic/citric acid-based NADES (< 10 g)	1*
Methanol (< 10 mL)	6
Instruments	
Centrifuge	1
Vortex	1
LC (≤ 1.5 kWh/sample)	1
Occupational hazard	0
Waste	6
	∑16*
Total penalty points	16
Analytical Eco-Scale total score	84*

*Same results were obtained for the NADES formed by lactic or citric acids. In the case of other NADES, no penalty points were assigned to NADES components, and the final score obtained was 85.

Analytical Greenness Calculator (AGREE)

(b)

1. Sample treatment (external and batch analysis)
2. Sample amount (1.0 mL)
3. Devise positioning (off-line)
4. Sample preparation stages (3 or fewer)
5. Automation, miniaturization (semi-automatic and miniaturized)
6. Derivatization (none)
7. Waste (1.160 mL)
8. Analysis throughput (12 analytes; 5 samples/hour)
9. Energy consumption (LC, most energy-intensive technique)
10. Source of reagents (some bio-based reagents)
11. Toxicity (Yes, methanol)
12. Operator's safety (highly flammable solvent)

Blue Applicability Grade Index (BAGI)

(c)

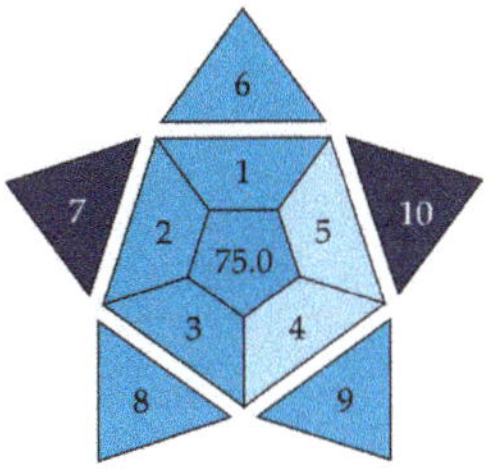

1. Type of analysis (quantitative)
2. Multi-element analysis (12 compounds)
3. Analytical technique (simple instrumentation available in most labs)
4. Simultaneous sample preparation (2-12 samples)
5. Sample preparation (miniaturized extraction)
6. Samples per hour (5 samples)
7. Reagents and materials (common commercially available reagents)
8. Preconcentration (required; one-step preconcentration)
9. Degree of automation (semi-automated with home-made systems)
10. Amount of sample (< 10 mL food/environmental)

Figure 5. Results obtained by the (**a**) Analytical Eco-Scale, (**b**) AGREE, and (**c**) BAGI metric tools of the developed NADES-LLME-nano-LC-UV method.

Finally, the functionality and applicability of the developed method, based on the blueness of the white analytical chemistry, was assessed using the BAGI tool [28]. Figure 5c shows the weak and strong points of the method in these terms in a blue asteroid-shaped pictogram. The final score of 75.0 out of 100 shows that the method is practical (recommended result is higher than 60 points). The same result was obtained for all NADES used, considering the easy commercial availability of NADES components and solvents. The main improvements to the method are the type of analysis that requires the use of a mass spectrometry system that allows the confirmation of the results obtained (point 1), the preparation of a higher number of simultaneous samples (point 4), and the miniaturization of extraction system versus the possibility of simpler systems or no sample preparation (point 5). Moreover, the use of common, commercially available reagents benefits simplicity relative to the method application (point 7). Therefore, despite these improvable elements of the developed methodology, in general terms, it is a simple and cheap analysis method.

4. Conclusions

For the first time, a method based on NADES combined with nano-LC-UV has been developed for the evaluation of alkylphenols and BPA in vegetable oil samples (olive and sunflower) as model compounds and matrices, respectively. The separation was studied in terms of stationary phase, gradient mode, mobile phase, and injection solvent compositions. The best results were obtained using an XBridge C_{18} capillary column, achieving a good separation in 11 min in step gradient mode. The injection solvent was based on a methanol/water mixture as the dilution solvent for hydrophilic NADES.

The green and blue profile of the proposed method was calculated to evaluate its environmental and practical scope. In this context, the Analytical Eco-Scale, the AGREE, and the BAGI metric tools returned scores of 84, 0.66, and 75.0, respectively. These data demonstrate good alignment with the green principles whilst showing good usability.

The proposed methodology, based on the results obtained, could be applied to the analysis of other polar and non-polar compounds in several oily samples. In addition, the improvement of the detection system for a more selective and sensible one could allow a better combination with NADES that present UV absorption, thus increasing the scope of application of these solvents. In this sense, the future path of DES must be directed toward a better understanding of their properties and how the components interact with each other, so that the compatibility of DES in different analytical processes can be improved.

Author Contributions: Conceptualization, Á.S.-M., G.D. and B.S.-R.; methodology, Á.S.-M. and G.D.; validation, Á.S.-M.; formal analysis, Á.S.-M.; investigation, Á.S.-M., G.D. and B.S.-R.; writing—original draft preparation, Á.S.-M., G.D., B.S.-R., S.F. and M.Á.R.-D.; supervision, G.D. and B.S.-R.; project administration, G.D. and M.Á.R.-D.; funding acquisition, G.D. and M.Á.R.-D. All authors have read and agreed to the published version of the manuscript.

Funding: This work was supported by the Spanish Ministry of Economy, Industry and Competitiveness (project AGL2017-89257-P) as well as by the University of La Laguna, and the Spanish Ministry of Science, Innovation and Universities, and the National Recovery and Resilience Plan (NRRP), Mission 4 Component 2 Investment 1.3—Call for proposals No. 341 of 15 March 2022 of Italian Ministry of University and Research funded by the European Union—NextGenerationEU; Award Number: Project code PE00000003, Concession Decree No. 1550 of 11 October 2022 adopted by the Italian Ministry of University and Research, CUP D93C22000890001, Project title "ON Foods—Research and innovation network on food and nutrition Sustainability, Safety and Security—Working ON Foods" (ONFOODS).

Data Availability Statement: Data are contained within the article.

Acknowledgments: Á.S.-M. would like to thank his FPI fellowship and his grant for short stays abroad at the CNR-ISB (Rome) cofinanced by Agencia Canaria de Investigación, Innovación y Sociedad de la Información de la Consejería de Economía, Conocimiento y Empleo y Fondo Social Europeo (FSE) Programa Operativo Integrado de Canarias 2014–2020, Eje 3 Tema Prioritario (85%).

Conflicts of Interest: The authors declare no conflict of interest.

References

1. United Nations A/RES/70/1. Resolution Adopted by the General Assembly on 25 September 2015. Transforming Our World: The 2030 Agenda for Sustainable Development. General Assembly of the United Nations 2015. Available online: https://undocs.org/en/A/RES/70/1 (accessed on 11 April 2024).
2. Nowak, P.M.; Wietecha-Posłuszny, R.; Pawliszyn, J. White Analytical Chemistry: An Approach to Reconcile the Principles of Green Analytical Chemistry and Functionality. *TrAC—Trends Anal. Chem.* **2021**, *138*, 116223. [CrossRef]
3. Ballesteros-Vivas, D.; Socas-Rodríguez, B.; Mendiola, J.A.; Ibáñez, E.; Cifuentes, A. Green Food Analysis: Current Trends and Perspectives. *Curr. Opin. Green Sustain. Chem.* **2021**, *31*, 100522. [CrossRef]
4. Koel, M. Do We Need Green Analytical Chemistry? *Green Chem.* **2016**, *18*, 923–931. [CrossRef]
5. Anastas, P.T.; Warner, J.C. *Green Chemistry: Theory and Practice*; Online; Oxford University Press: Oxford, UK, 2000; ISBN 9780197704059.
6. Fanali, S.; D'Orazio, G.; Quaglia, M.G. Separation of Aryl Propionic Acids by Capillary Liquid Chromatography. *Chromatographia* **2004**, *60*, S239–S243. [CrossRef]

7. Liu, W.; Zhang, K.; Chen, J.; Yu, J. Ascorbic Acid and Choline Chloride: A New Natural Deep Eutectic Solvent for Extracting Tert-Butylhydroquinone Antioxidant. *J. Mol. Liq.* **2018**, *260*, 173–179. [CrossRef]
8. Asensio-ramos, M.; Fanali, C.; D'Orazio, G.; Fanali, S. Nano-Liquid Chromatography. In *Liquid Chromatography: Fundamentals and Instrumentation*; Fanali, S., Haddad, P.R., Poole, C.F., Riekkola, M.-L., Eds.; Elsevier: Amsterdam, The Netherlands, 2017; pp. 637–695. ISBN 9780128053935.
9. dos Santos, N.G.P.; Maciel, E.V.S.; Medina, D.A.V.; Lanças, F.M. NanoLC-EI-MS: Perspectives in Biochemical Analysis. *Int. J. Mol. Sci.* **2023**, *24*, 11746. [CrossRef]
10. Cestaro, B.I.; Machado, K.C.; Batista, M.; da Silva, B.J.G. Hollow-Fiber Liquid Phase Microextraction for Determination of Fluoxetine in Human Serum by Nano-Liquid Chromatography Coupled to High Resolution Mass Spectrometry. *J. Chromatogr. B* **2024**, *1234*, 124018. [CrossRef]
11. Perna, F.M.; Vitale, P.; Capriati, V. Deep Eutectic Solvents and Their Applications as Green Solvents. *Curr. Opin. Green Sustain. Chem.* **2020**, *21*, 27–33. [CrossRef]
12. Prabhune, A.; Dey, R. Green and Sustainable Solvents of the Future: Deep Eutectic Solvents. *J. Mol. Liq.* **2023**, *379*, 121676. [CrossRef]
13. Álvarez, M.S.; Longo, M.A.; Rodríguez, A.; Deive, F.J. The Role of Deep Eutectic Solvents in Catalysis. A Vision on Their Contribution to Homogeneous, Heterogeneous and Electrocatalytic Processes. *J. Ind. Eng. Chem.* **2024**, *132*, 36–49. [CrossRef]
14. Al-Farsi, R.; Hayyan, M. Paving the Way for Advancement of Renewable Energy Technologies Using Deep Eutectic Solvents: A Review. *Renew. Sustain. Energy Rev.* **2023**, *184*, 113505. [CrossRef]
15. Jafari, K.; Fatemi, M.H.; Estellé, P. Deep Eutectic Solvents (DESs): A Short Overview of the Thermophysical Properties and Current Use as Base Fluid for Heat Transfer Nanofluids. *J. Mol. Liq.* **2021**, *321*, 114752. [CrossRef]
16. Ge, X.; Gu, C.; Wang, X.; Tu, J. Deep Eutectic Solvents (DESs)-Derived Advanced Functional Materials for Energy and Environmental Applications: Challenges, Opportunities, and Future Vision. *J. Mater. Chem. A* **2017**, *5*, 8209–8229. [CrossRef]
17. Mannucci, G.; Teixeira, G.; Sosa, F.H.B.; Palluzzi, M.; Busato, M.; Coutinho, J.A.P.; D'Angelo, P. Predicting the Thermal Behavior in the Design of Type V Deep Eutectic Solvents: The Combined Role of Polarity and Steric Asymmetry. *ACS Sustain. Chem. Eng.* **2024**, *12*, 2862–2870. [CrossRef]
18. Santana-Mayor, Á.; Rodríguez-ramos, R.; Herrera-herrera, A.V.; Socas-Rodríguez, B.; Rodríguez-Delgado, M.Á. Trends in Analytical Chemistry Deep Eutectic Solvents. The New Generation of Green Solvents in Analytical Chemistry Alvaro. *TrAC—Trends Anal. Chem.* **2021**, *134*, 116108. [CrossRef]
19. Cunha, S.C.; Lapa, R.; Fernandes, O. Trends in Analytical Chemistry Deep Eutectic Solvents for Microextraction of Endocrine Disruptor Contaminants from Water and Food: Review. *TrAC—Trends Anal. Chem.* **2024**, *171*, 117495. [CrossRef]
20. Mansour, F.R.; Bedair, A.; Hamed, M.; Magdy, G.; Ali, I.; Locatelli, M. Applications of (Natural) Deep Eutectic Solvents in Liquid Phase Microextraction: A Review. *Microchem. J.* **2024**, *198*, 110178. [CrossRef]
21. Molnar, M.; Gašo-Sokač, D.; Komar, M.; Jakovljević Kovač, M.; Bušić, V. Potential of Deep Eutectic Solvents in the Extraction of Organic Compounds from Food Industry By-Products and Agro-Industrial Waste. *Separations* **2024**, *11*, 35. [CrossRef]
22. Fanali, S.; Aturki, Z.; D'Orazio, G.; Rocco, A.; Ferranti, A.; Mercolini, L.; Raggi, M.A. Analysis of Aloe-Based Phytotherapeutic Products by Using Nano-LC-MS. *J. Sep. Sci.* **2010**, *33*, 2663–2670. [CrossRef] [PubMed]
23. Khezeli, T.; Daneshfar, A.; Sahraei, R. A Green Ultrasonic-Assisted Liquid-Liquid Microextraction Based on Deep Eutectic Solvent for the HPLC-UV Determination of Ferulic, Caffeic and Cinnamic Acid from Olive, Almond, Sesame and Cinnamon Oil. *Talanta* **2016**, *150*, 577–585. [CrossRef]
24. Sang, J.; Li, B.; Huang, Y.Y.; Ma, Q.; Liu, K.; Li, C.Q. Deep Eutectic Solvent-Based Extraction Coupled with Green Two-Dimensional HPLC-DAD-ESI-MS/MS for the Determination of Anthocyanins from *Lycium ruthenicum* Murr. Fruit. *Anal. Methods* **2018**, *10*, 1247–1257. [CrossRef]
25. Huang, Y.; Feng, F.; Jiang, J.; Qiao, Y.; Wu, T.; Voglmeir, J.; Chen, Z.G. Green and Efficient Extraction of Rutin from Tartary Buckwheat Hull by Using Natural Deep Eutectic Solvents. *Food Chem.* **2017**, *221*, 1400–1405. [CrossRef]
26. Gałuszka, A.; Migaszewski, Z.M.; Konieczka, P.; Namieśnik, J. Analytical Eco-Scale for Assessing the Greenness of Analytical Procedures. *TrAC—Trends Anal. Chem.* **2012**, *37*, 61–72. [CrossRef]
27. Pena-Pereira, F.; Wojnowski, W.; Tobiszewski, M. AGREE—Analytical GREEnness Metric Approach and Software. *Anal. Chem.* **2020**, *92*, 10076–10082. [CrossRef]
28. Manousi, N.; Wojnowski, W.; Płotka-Wasylka, J.; Samanidou, V. Blue Applicability Grade Index (BAGI) and Software: A New Tool for the Evaluation of Method Practicality. *Green Chem.* **2023**, *25*, 7598–7604. [CrossRef]
29. National Library of Medicine (US). National Center for Biotechnology Information PubChem. Available online: https://pubchem.ncbi.nlm.nih.gov/ (accessed on 8 March 2024).
30. Kennedy, R.T.; Jorgenson, J.W. Preparation and Evaluation of Packed Capillary Liquid Chromatography Columns with Inner Diameters from 20 to 50 Mm. *Anal. Chem.* **1989**, *61*, 1128–1135. [CrossRef]
31. Vissers, J.P.C. Recent Developments in Microcolumn Liquid Chromatography. *J. Chromatogr. A* **1999**, *856*, 117–143. [CrossRef]
32. Vissers, J.P.C.; De Ru, A.H.; Ursem, M.; Chervet, J.P. Optimised Injection Techniques for Micro and Capillary Liquid Chromatography. *J. Chromatogr. A* **1996**, *746*, 1–7. [CrossRef]
33. Héron, S.; Tchapla, A.; Chervet, J.P. Influence of Injection Parameters on Column Performance in Nanoscale High-Performance Liquid Chromatography. *Chromatographia* **2000**, *51*, 495–499. [CrossRef]

34. Rodríguez-Ramos, R.; Herrera-Herrera, A.V.; Díaz-Romero, C.; Socas-Rodríguez, B.; Rodríguez-Delgado, M.Á. Eco-Friendly Approach Developed for the Microextraction of Xenobiotic Contaminants from Tropical Beverages Using a Camphor-Based Natural Hydrophobic Deep Eutectic Solvent. *Talanta* **2024**, *266*, 124932. [CrossRef]
35. Baute-Pérez, D.; Santana-Mayor, Á.; Herrera-Herrera, A.V.; Socas-Rodríguez, B.; Rodríguez-Delgado, M.Á. Analysis of Alkylphenols, Bisphenols and Alkylphenol Ethoxylates in Microbial-Fermented Functional Beverages and Bottled Water: Optimization of a Dispersive Liquid-Liquid Microextraction Protocol Based on Natural Hydrophobic Deep Eutectic Solvents. *Food Chem.* **2022**, *377*, 131921. [CrossRef]
36. Wang, Y.; Li, J.; Sun, D.; Yang, S.; Liu, H.; Chen, L. Strategies of Dispersive Liquid-Liquid Microextraction for Coastal Zone Environmental Pollutant Determination. *J. Chromatogr. A* **2021**, *1658*, 462615. [CrossRef]
37. Santana-Mayor, Á.; Socas-Rodríguez, B.; Rodríguez-Ramos, R.; Rodríguez-Delgado, M.Á. A Green and Simple Procedure Based on Deep Eutectic Solvents for the Extraction of Phthalates from Beverages. *Food Chem.* **2020**, *312*, 125798. [CrossRef]
38. Santana-Mayor, Á.; Rodríguez-Ramos, R.; Socas-Rodríguez, B.; Rodríguez-Delgado, M.Á.; D'Orazio, G. Nano-Liquid Chromatography Combined with a Sustainable Microextraction Based on Natural Deep Eutectic Solvents for Analysis of Phthalate Esters. *Electrophoresis* **2020**, *41*, 1768–1775. [CrossRef]
39. Rodríguez-Juan, E.; Rodríguez-Romero, C.; Fernández-Bolaños, J.; Florido, M.C.; Garcia-Borrego, A. Phenolic Compounds from Virgin Olive Oil Obtained by Natural Deep Eutectic Solvent (NADES): Effect of the Extraction and Recovery Conditions. *J. Food Sci. Technol.* **2021**, *58*, 552–561. [CrossRef]
40. Jing, X.; Cao, C.; Wu, W.; Zhao, W.; Wang, Y. Deep Eutectic Solvent-Based Vortex-Assisted Dispersive Liquid–Liquid Microextraction Combined with High Performance Liquid Chromatography for the Determination of Phenolic Acids in Vegetable Oils. *Food Sci. Technol. Res.* **2019**, *25*, 563–568. [CrossRef]
41. Coscarella, M.; Nardi, M.; Alipieva, K.; Bonacci, S.; Popova, M.; Procopio, A.; Scarpelli, R.; Simeonov, S. Alternative Assisted Extraction Methods of Phenolic Compounds Using NaDESs. *Antioxidants* **2024**, *13*, 62. [CrossRef]
42. National Water Quality Monitoring Council; U.S. Geological Survey (USGS); (USEPA), U.S.E.P.A. NEMI—National Environamental Methods Index. Available online: https://www.nemi.gov/home/ (accessed on 8 March 2024).
43. Płotka-Wasylka, J. A New Tool for the Evaluation of the Analytical Procedure: Green Analytical Procedure Index. *Talanta* **2018**, *181*, 204–209. [CrossRef]
44. Płotka-Wasylka, J.; Wojnowski, W. Complementary Green Analytical Procedure Index (ComplexGAPI) and Software. *Green Chem.* **2021**, *23*, 8657–8665. [CrossRef]
45. Wojnowski, W.; Tobiszewski, M.; Pena-pereira, F.; Psillakis, E. AGREEprep—Analytical Greenness Metric for Sample Preparation. *TrAC—Trends Anal. Chem. Anal. Chem.* **2022**, *149*, 116553. [CrossRef]

separations

Article

Automated Image Analysis for Retention Determination in Centrifugal Partition Chromatography

Felix Buthmann, Florian Pley, Gerhard Schembecker * and Jörg Koop

Laboratory of Plant and Process Design, Department of Biochemical and Chemical Engineering, TU Dortmund University, 44227 Dortmund, Germany
* Correspondence: gerhard.schembecker@tu-dortmund.de; Tel.: +49-(0)231-755-2339

Abstract: In Centrifugal Partition Chromatography, two immiscible liquids are used as mobile and stationary phases. During operation, bleeding of the stationary phase cannot be eliminated completely. For optimal separation performance, however, it is crucial to maintain sufficient amounts of stationary phase in the system, which is quantitatively measured by the retention value. With an online measurement of that retention value, it is possible to make predictions about the separation performance of the system. Therefore, an image processing algorithm was developed in this study, allowing quick and effortless online evaluation of retention by image analysis. Finally, the results were compared with proven analysis methods to evaluate the measurements' validity. With the help of the new algorithm, it was possible to improve the number of pictures analyzed per time and the precision compared to the previously used technique.

Keywords: Centrifugal Partition Chromatography; flow visualization; stationary phase retention; automated evaluation; image processing algorithm

1. Introduction

Countercurrent Chromatography can be differentiated into two different technical concepts: instruments containing at least one coil that rotates in a planetary trajectory (hydrodynamic CCC) and apparatuses consisting of rotor discs with embedded cells rotating around a single axis (hydrostatic CCC). The latter is also known as Centrifugal Partition Chromatography (CPC) and is particularly interesting for industrial applications due to its comparatively straightforward technical implementation. Although the term Countercurrent Chromatography (CCC) suggests otherwise, one liquid phase is immobilized in such an apparatus being the stationary phase, and a second liquid phase is actively pumped as a mobile phase. Thus, when compared to conventional liquid chromatography, the solid stationary phase is replaced by a second liquid immiscible with the mobile phase that is kept in the CPC apparatus by centrifugal force [1].

The use of two liquid phases has several advantages over a solid-liquid system, for example, high loading capacities and total recovery since no irreversible adsorption is possible in a liquid-liquid system. In terms of costs per capacity, liquid stationary phases are usually much less expensive than packed solids. Furthermore, the packing process of the solid material is no longer a source of error, and environmentally friendly solvents (e.g., Aqueous Two-Phase Systems) are available. At the same time, solvents are ideal for processing sensitive substances such as proteins [1–3]. Besides, there are further fields of application: In the context of Quantitative Structure–Activity Relationship (QSAR) analysis, CPC can be used to determine partition coefficients. With the help of these data sets, QSAR models can be implemented, allowing for the prediction of partition coefficients based on the molecular structure [4,5].

For initial start-up, the centrifugal chromatograph is filled with a stationary phase. The mobile phase is subsequently pumped into the rotor and is dispersed within the stationary phase at each chamber inlet. The mass transfer takes place in the resulting dispersion zone. A coalescence zone of the mobile phase is formed at each chamber outlet. Thus, in an ideal case, each CPC chamber acts as a mixer-settler unit known for extraction processes. In order to work as described, retention is a key parameter to ensure optimal separation performance. Until today, the determination of that crucial parameter requires manual work and time after each experimental CPC run. Therefore, quantitative online monitoring of the retention was hardly performable, and the total number of parameter sets being investigated according to their retention was severely limited [6]. With rapidly evolving image processing technology and the resulting potentials, the goals of the work reported here are, therefore, to reduce the duration of the evaluation on the one hand and to increase the accuracy on the other hand [7]. For this reason, a Matlab-based routine was developed and implemented, automating the analysis of raw data supplied by a high-speed camera [8]. Image processing algorithms were used to eliminate the need for manual editing, speeding up the determination of retention as well as removing the human error source from the process.

2. Theory

For the determination of retention in CPC, a visual inspection ensures precise raw data. Therefore, it is crucial to utilize transparent rotors. The combination of camera technology and sophisticated image analysis has already been used to obtain information about the flow regime present [9,10]. In this study, the determination of the retention value Sf^* (Equation (1)) is of particular interest [6].

$$Sf^* = \frac{V_{stat,c} + V_{stat,d}}{V_c} \tag{1}$$

The Sf^* is calculated from the volume of stationary phase inside the chambers $V_{stat,c}$ and ducts $V_{stat,d}$ as well as the total volume of the chambers V_c and provides information on the amount of stationary phase remaining in each chamber during operation. The optimum range is between $0.25 \leq Sf \leq 0.75$, with low Sf^* values resulting in a chromatogram showing short retention times and closely spaced peaks. Identical components elute much later and with higher resolution at high Sf^* values [1].

In order to determine Sf^*, mobile phase flow has to be stopped temporarily. Next, a video of the resting liquid phases in the chambers and ducts can be recorded. It is a state of the art to process the raw data by hand [6]. For this purpose, a polygonal path is placed around a single chamber-duct combination. This procedure has to be repeated for each chamber, from which an Sf^* shall be calculated. A manually marked polygonal path is shown in Figure 1 as an example, where the stationary phase occupies part of the chamber as well as part of the channel.

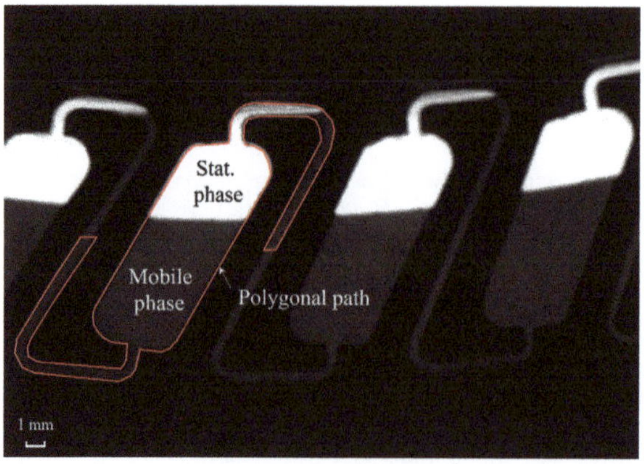

Figure 1. Polygonal path (red) highlighting a single chamber-duct combination.

The number of white pixels within this area is then evaluated using the image processing tool ImageJ [11]. The actual *Sf** is determined subsequently by calculating the ratio of bright pixels to the total number of pixels inside the chamber [6].

3. Materials and Methods

3.1. Retention: Raw Data Acquisition

3.1.1. Chemicals

For all experiments performed during this study, the Arizona N system was used [12]. It was prepared by adding equal volumes of ethyl acetate (purity = 99.9%), heptane (purity = 99.8%), methanol (purity = 99%), all purchased from VWR International (Radnor, PA, USA), and water purified by a MILLI-Q® system (Millipak® Express 40, Merck, Darmstadt, Germany). The aqueous phase was selectively dyed (Methylene blue, Merck KGaA, 20 mg/L).

3.1.2. Rotor Configuration and Manual *Sf** Determination

A semi-transparent rotor was used in a modified CPC (Type: FCPC, Chromaton® (Annonay, France), descending mode, 1000 rpm, 15 mL/min mobile phase flow) as it has already been described in the literature [5,6]. The manual retention evaluation method was executed as described in Section 2. The rotor consists of a single disk with 66 chambers and connecting channels embedded. In contrast to stacked rotor versions, which are used for actual separation tasks (stacked and interconnected individual discs ensure a higher number of theoretical stages), the rotor mentioned is suitable for visual fluid flow detection.

3.2. Newly Developed Image Processing Algorithm

The raw data for the new algorithm were still images extracted from the high-speed footage recorded at defined time intervals t_{ext} (Figure 2A). The edge detection algorithm "Canny" was then applied to the gray-scale images. It is based on the search for local maxima of the intensity gradient, where the latter is determined by smoothing the image with a Gaussian filter (Figure 2B) [13].

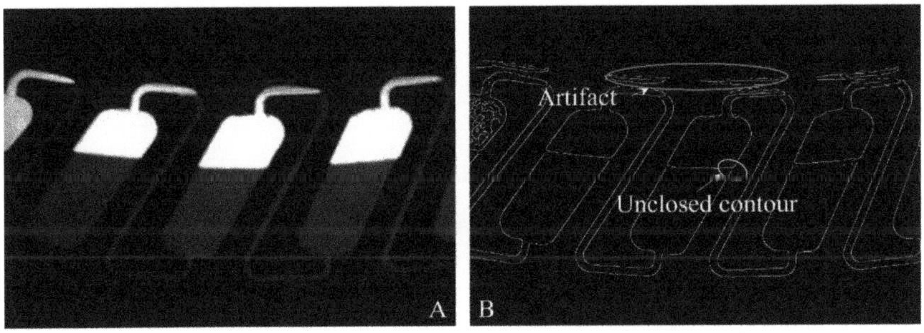

Figure 2. (**A**) Unprocessed still image—extracted from video files of the high-speed camera. (**B**) Binary image after editing with Canny edge algorithm. Highlighted: Artifact due to reflections (red) and unclosed contour (yellow).

The binary image generated was examined for unclosed contours, according to Kovesi, and those gaps were filled [14]. For this purpose, a circular area with a radius of half the width of the gap to be closed (user-defined parameter) was created at the corner points of each edge segment. If two of these circles overlapped, they were reduced to a one-pixel-wide direct connection. On the other hand, the original edge was restored if there were no contact points. Besides the correction of unclosed contours, it was also necessary to eliminate artifacts in the binary image. These were caused by camera flash reflections in the viewing windows' aluminum framing, which would interfere with further image processing.

In order to do so, a cutout of a ring-shaped mask was multiplied by the existing binary image (Figure 3A). This was possible because the data formats are compatible: multiplying a matrix with binary values to another matrix with the same data type gives the intersection of positive data points (pixels).

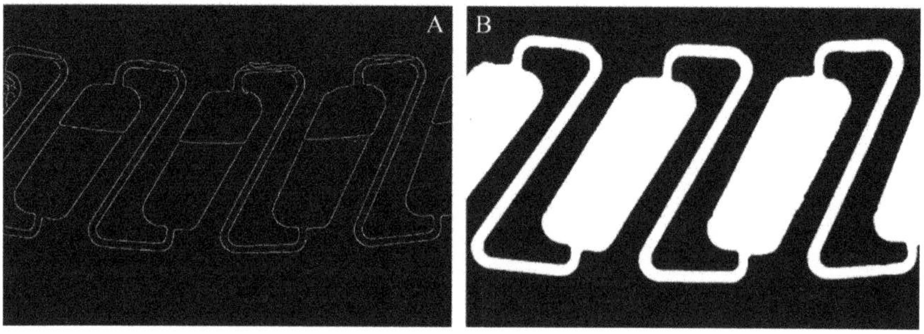

Figure 3. (**A**) Binary image after filtering with the help of a circular mask and edge closure, according to Kovesi. (**B**) Enclosed areas are filled.

The areas enclosed by contours were then set to the binary value 1, which is why they appear white (Figure 3B). To calculate the retention value, Sf^* (Equation (1)) was subdivided into a proportion assigned to the chambers (Sf^*_c) and a proportion assigned to the ducts (Sf^*_d), as the stationary phase occupies part of the chamber and part of the duct (Figure 1). Besides, the result was normalized to the corresponding number of chambers n_c and ducts n_d evaluated (Equation (2)).

$$Sf_c^* = \frac{\frac{V_{stat,c}}{n_c}}{\frac{V_c}{n_c}} \; Sf_d^* = \frac{\frac{V_{stat,d}}{n_d}}{\frac{V_c}{n_c}} \tag{2}$$

To determine the variables n_d, n_c, $V_{stat,c}$, $V_{stat,d}$, and V_c information from the unprocessed still image (Figure 2A) and the binary image (Figure 3B) was used. The volume of the chambers (V_c) and ducts were calculated by multiplying their depth in the direction of the rotation axis with their surface area. The surface area and the number of chambers and ducts (n_d, n_c) were determined by isolating the chambers through a sequence of erode and dilate commands (Figure 4A). Another binary image with exposed ducts was obtained analogously by subtracting the result obtained from the aforementioned binary image (Figure 4B). The unknown number of ducts and chambers and their surface area/volume were evaluated with this sequence of operations.

The fraction of stationary phase in the chambers and ducts was determined by using each chamber or duct isolated as a mask and multiplying it with the extracted non-binary still image. After that, the images obtained were binarized again (Figure 4C,D). The proportion of white pixels was evaluated and provided information about the remaining variables $V_{stat,c}$ and $V_{stat,d}$. In this context, it was essential to consider which phase was colored and therefore appears dark in the video material. In the example presented in Figure 2, the heavy phase was dyed, and the centrifugal chromatograph was operated in descending mode. However, it is also possible to analyze in *ascending mode* or when the light phase is dyed (user parameters).

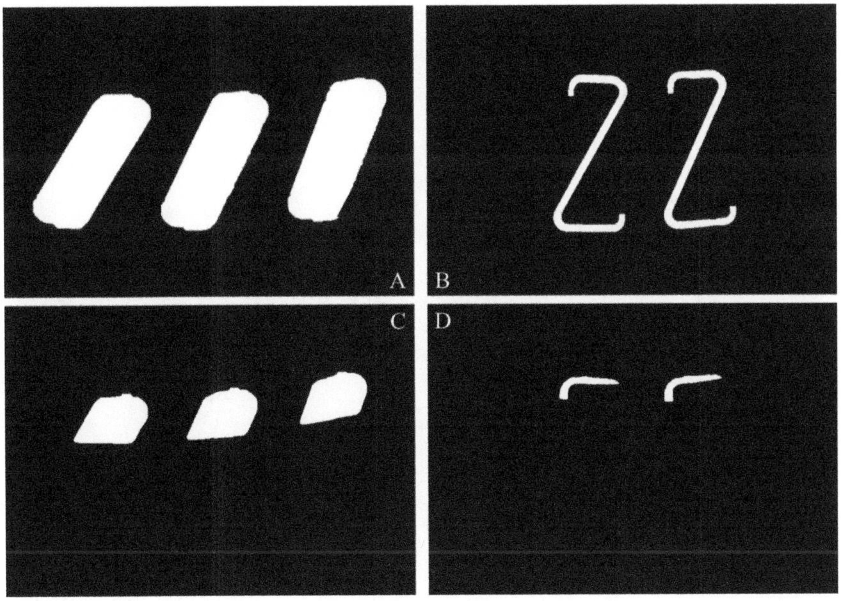

Figure 4. (**A**) Binary image with isolated chambers. (**B**) Binary image with isolated ducts. (**C**) Binary image showing the proportion of non-dyed fluid within chambers. (**D**) Binary image showing the proportion of non-dyed fluid within ducts.

3.3. Method Comparison

For comparison of the manual and automated measurement methods, the Bland–Altman plot was utilized. [15] When a sample with i different data points is analyzed, two different methods give 2i measurement results. The data points can be plotted with the mean of the two methods on the abscissa and the respective difference on the ordinate, resulting in the Bland–Altman plot with i data points. Additional lines are often plotted for visual support, representing the 95% confidence interval. In combination, this information can be used to identify any differences (e.g., proportional bias) between the two measurement methods.

4. Results and Discussion

Benchmarking the Image Analysis Routine

The data set recorded according to Section 3.1.2 was analyzed by applying the following user parameters: descending mode, heavy phase dyed, t_{ext} = 0.5 s, resulting in 18 still images for each of the six videos recorded. For those images, a value for the retention of the stationary phase in the chambers, the ducts, and the chamber-duct combinations was determined using the procedure described in the previous section. The value for the chamber-duct-combination was re-evaluated via the manual image analysis described above, serving as a benchmark (Section 2). A mean retention value of 0.2921 was obtained for the manual method, and the evaluation via edge detection resulted in a mean value of 0.2916. That similarity between the mean values is a necessary but not yet a sufficient condition for the equivalence of both measurement methods. Considering the standard deviations of retention values for single inspection windows, a significant difference in precision is noticeable (Figure 5).

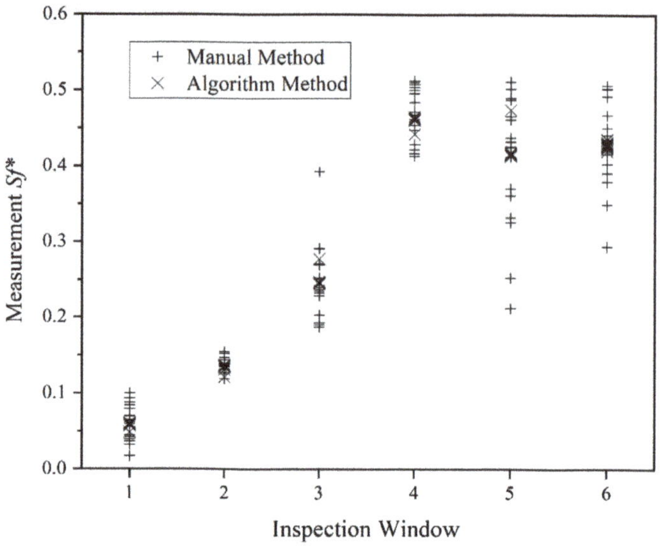

Figure 5. Retention values of the six inspection windows of the rotor. Results processed with the automated method are indicated with crosses, plus signs represent manually determined retention values.

The mean retention value for the first inspection window is 0.0584 when measured with the automated method and 0.0611 using the manual method. However, the standard deviation differs tenfold, being 0.0027 with the automated and 0.0228 with the manual method. This trend is present for all inspection windows and indicates the superiority of the new image analysis routine developed.

A simple Pearson's Correlation would not be sufficient to compare those paired data sets further. However, the Bland–Altman plot is appropriate for this purpose [15]. As mentioned, the same dataset (raw data available as Supplementary Materials) of still images was analyzed with two methods for direct method comparison. Furthermore, according to Bland and Altman, a graphical evaluation yields in the plot shown in Figure 6.

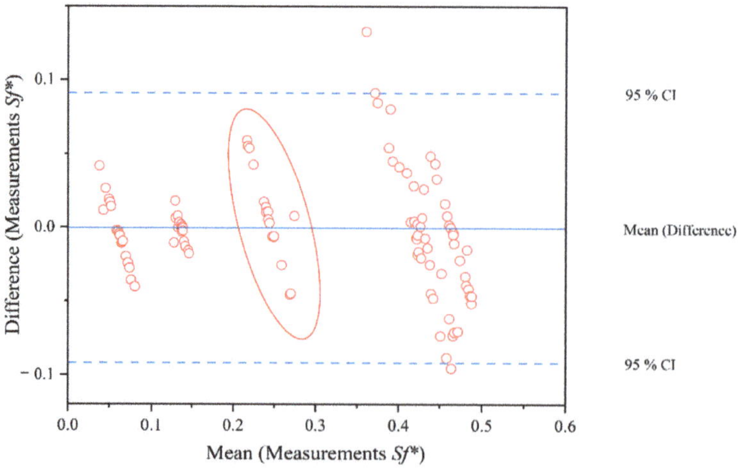

Figure 6. Bland–Altman plot comparing the automated image analysis with the manual variant.

The relative difference of the retention values derived from the two measurement methods is plotted over the mean Sf^*-value of both. It is noticeable that the variability of the evaluation methods for retention values in the range $Sf^* \leq 0.3$ is small (with values of $+/- 0.06$) and also lies within the 95% confidence interval (blue dashed lines) in all cases.

With higher mean Sf^* values, the relative difference between the methods increases, whereby with Sf^* values > 0.3, the 95 % confidence interval is exceeded in a few cases. Nevertheless, no significant systematic error can be identified, indicating an inconsistency in the measurement methods being compared. Considering Figure 5, the automated evaluation can be referred to as the new "gold standard" because of its significantly lower standard deviation. The angled cluster of data points for every inspection window (Figure 6, red ellipse marking the third inspection window as an example) is another indicator of the lower accuracy of the manual evaluation method. Such a pattern occurs for a collection of data points with a similar true value if one method's precision is higher. For instance, if the manual method gives a value too low, while the automated method gives a value close to the true value, the mean value is shifted left, and the relative difference is shifted up (Section 3.3) and vice versa. However, if both methods show similar deviations from the true value, angled clusters like those in Figure 6 do not occur.

In summary, the average discrepancy between the methods is below 1%, which can be considered negligible. Besides, there is no proportional bias evident. It can be stated that the measurement methods compared in this section provide the same accuracy, with the automated method providing more precise data. Apart from precision, there are other criteria for comparing both techniques. The amount of time saved is an argument favoring automated edge detection. The effort for processing the data given (108 still images) is cut from several hours to just 55 s. Another important aspect is the exclusion of human error sources. While there is an observable wandering deviation with changing personnel using the manual method, the automated analysis ensures reproducible results. In addition, the automated evaluation provides the possibility for retention determination during operation.

5. Conclusions

Using fully automated image processing, we developed a new routine for analyzing retention inside a Centrifugal Partition Chromatograph. After comparison with the manual evaluation of the raw data with the help of a Bland–Altman plot, we demonstrated that both methods provide equivalently accurate results. The main advantages of automated evaluation, the lower standard deviation combined with significant time saving, allow for new process monitoring and control potentials. The determination of locally resolved retention values in the entire rotor during operation is now ready to be implemented, while the vast number of images is almost impossible to process manually. Identifying the current operating status of a centrifugal chromatograph is a huge step towards the implementation of those novel apparatuses in industry. Furthermore, the data can be used to control the operating parameters of the apparatus to prevent progressive flooding with the mobile phase (so-called bleeding), enabling long-term operability.

Of course, the rotor described is not particularly suitable for separation tasks. As mentioned, rotors in this context comprise more chambers to ensure sufficient separation performance. The rotor used in this study, which is observable with the camera technique, has a crucial function in understanding the influence of operating conditions and design parameters. After the interconnections are understood, the knowledge gained can be transferred to rotors consisting of several non-transparent disks. Compared with different approaches to measure the retention in Centrifugal Partition Chromatography (for example, phase separation at the device outlet and subsequent ratio determination of mobile and stationary phase), the innovative approach presented enables local resolution of bleeding within the rotor. Thus, the exact location of discharged stationary phase, and consequently, local hydrodynamic causes, can be identified. For instance, that is important when investigating the impact of sample injection. Because the sample contains at least one additional component compared to the mobile phase, it is not in equilibrium with the stationary phase.

Therefore, it may initiate stationary sample loss, although the CPC was in stable operation before. In the next step, it is planned to analyze bleeding in a temporally as well as spatially resolved manner.

Supplementary Materials: The following supporting information can be downloaded at: https://www.mdpi.com/article/10.3390/separations9110358/s1. File S1: Raw data.

Author Contributions: Conceptualization: F.B., G.S. and J.K.; methodology: F.B.; software: F.B. and F.P.; validation: F.B. and J.K.; formal analysis: F.B.; investigation: F.B. and F.P.; data curation: F.B.; writing—original draft preparation: F.B.; writing—review and editing: F.B., G.S. and J.K.; visualization: F.B.; supervision: G.S. and J.K. All authors have read and agreed to the published version of the manuscript.

Funding: This research received no external funding.

Data Availability Statement: Not applicable.

Acknowledgments: The authors would like to thank Daniela Ermeling and Frank Kernbach for their technical support.

Conflicts of Interest: The authors declare no conflict of interest.

References

1. Berthod, A. *Countercurrent Chromatography: The Support-Free Liquid Stationary Phase*; Elsevier: Amsterdam, The Netherlands, 2002; ISBN 9780444507372.
2. Friesen, J.B.; McAlpine, J.B.; Chen, S.-N.; Pauli, G.F. Countercurrent Separation of Natural Products: An Update. *J. Nat. Prod.* **2015**, *78*, 1765–1796. [CrossRef]
3. Szekely, G.; Zhao, D. *Sustainable Separation Engineering: Materials, Techniques and Process Development*; John Wiley & Sons Ltd.: Hoboken, NJ, USA, 2022; ISBN 9781119740087.
4. Kukula-Koch, W.; Kruk-Słomka, M.; Stępnik, K.; Szalak, R.; Biała, G. The Evaluation of Pro-Cognitive and Antiamnestic Properties of Berberine and Magnoflorine Isolated from Barberry Species by Centrifugal Partition Chromatography (CPC), in Relation to QSAR Modelling. *Int. J. Mol. Sci.* **2017**, *18*, 2511. [CrossRef]
5. Marsden-Jones, S.C. The Application of Quantitative Structure Activity Relationship Models to the Method Development of Countercurrent Chromatography. Ph.D. Thesis, Brunel University, London, UK, 2015.
6. Fromme, A. Systematic Approach Towards Solvent System Selection for Ideal Fluid Dynamics in Centrifugal Partition Chromatography. Master's Thesis, Technical University Dortmund, Dortmund, Germany, 2020.
7. Lins, J.; Harweg, T.; Weichert, F.; Wohlgemuth, K. Potential of Deep Learning Methods for Deep Level Particle Characterization in Crystallization. *Appl. Sci.* **2022**, *12*, 2465. [CrossRef]
8. Moler, C. *Matlab*; Mathworks: Natick, MA, USA, 1984.
9. Schwienheer, C.; Merz, J.; Schembecker, G. Investigation, comparison and design of chambers used in centrifugal partition chromatography on the basis of flow pattern and separation experiments. *J. Chromatogr. A* **2015**, *1390*, 39–49. [PubMed]
10. Fromme, A.; Fischer, C.; Keine, K.; Schembecker, G. Characterization and correlation of mobile phase dispersion of aqueous-organic solvent systems in centrifugal partition chromatography. *J. Chromatogr. A* **2020**, *1620*, 460990. [PubMed]
11. Rasband, W.S. *ImageJ*; U.S. National Institutes of Health: Bethesda, MD, USA, 1997.
12. Oka, F.; Oka, H.; Ito, Y. Systematic search for suitable two-phase solvent systems for high-speed counter-current chromatography. *J Chromatogr. A* **1991**, *538*, 99–108. [CrossRef]
13. Canny, J. A Computational Approach to Edge Detection. *IEEE T Pattern Anal.* **1986**, *PAMI-8*, 679–698. [CrossRef]
14. Kovesi, P. Functions for Computer Vision. Available online: https://www.peterkovesi.com/matlabfns/ (accessed on 15 August 2021).
15. Bland, J.M.; Altman, D.G. Statistical methods for assessing agreement between two methods of clinical measurement. *Lancet* **1986**, *1*, 307–310. [CrossRef]

separations

MDPI

Review

Sulfolane Analysis in Environmental Samples: A Critical Review

Merrik Kobarfard and Tadeusz Górecki *

Department of Chemistry, University of Waterloo, 200 University Avenue West, Waterloo, ON N2L 3G1, Canada; merrik.kobarfard@uwaterloo.ca
* Correspondence: tgorecki@uwaterloo.ca

Abstract: Sulfolane, a highly water-soluble industrial solvent, has raised environmental concerns due to its widespread usage and its prolonged persistence once released into the environment. To assess the extent of contamination effectively, reliable analytical methods are essential. In this review article, the existing methodologies for the sample preparation and determination of sulfolane are systematically examined and evaluated. In brief, existing guidelines for sampling from environmental matrices provide a solid foundation for sulfolane analysis. Notably, there is little variation in the choice of final determination methods, with GC-MS or GC-FID being favored across all studies. However, substantial variability emerges in sample preparation methods. Moreover, upon scrutiny of the analytical procedures utilized in sulfolane analysis, it becomes evident that a considerable number of them are characterized by inefficiency or environmental hazards, often relying on substantial quantities of chlorinated solvents. Through this review, valuable insights into the challenges at hand and potential solutions can be gained, offering a foundation for the development of novel sulfolane analysis methods applicable to a range of environmental matrices. Overall, this field calls for further research to devise efficient and environmentally sustainable analytical methods for sulfolane analysis.

Keywords: sulfolane analysis; environmental analysis; residue analysis; sulfolane contamination

1. Introduction

Sulfolane, also referred to as tetramethylene sulfone and tetrahydrothiophene 1,1-dioxide, is a highly versatile industrial solvent with the molecular formula $C_4H_8SO_2$ (Figure 1) [1]. Although first mentioned in the literature in 1916, it was not until the 1960s that it became commercially available. Since then, its usage has steadily increased, and it has found a wide range of applications in various fields [1]. The evidence lies in the tenfold increase in publications over the past two decades. While a fraction of these articles delve into the environmental impact, remediation, synthesis pathways, or analytical aspects of sulfolane, the majority are dedicated to unveiling its newfound applications. This rise in popularity can be attributed in part to its unique physical and chemical properties (Table 1). These properties include high boiling point; thermal stability; being miscible with a wide range of liquids including water, toluene, acetone, alcohols, and aromatic solvents; but not being dissolved in alkanes [2].

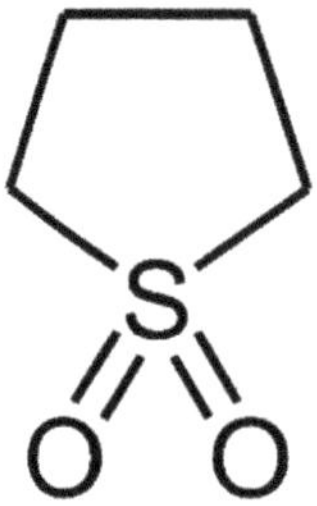

Figure 1. Structure of sulfolane.

Table 1. Chemical and physical properties of sulfolane.

Property	Value	Ref
Molecular formula	$C_4H_8SO_2$	[1]
Molecular weight	120.17 g/mol	[1]
Melting point	28.5 °C	[1]
Boiling point	287.3 °C	[1]
Log P	−0.4	[3]
Solubility in water	Miscible at 30 °C	[4]
pK_a	12.9 *	[5]

* Despite extensive research, there is only one reported pKa value for sulfolane. However, this value seems questionable given that dimethylsulfoxide has a pKa value of 35.1 and shares structural similarities with sulfolane [6].

Sulfolane has become a desirable solvent for many applications in different fields [1]. In the oil industry, sulfolane is used as an extractive solvent for the separation of benzene, toluene, and xylene from aliphatic hydrocarbons, while in the gas treatment processes, it is effective in removing acidic components such as hydrogen sulfide and carbon dioxide from sour natural gas, particularly in the western parts of Canada due to high concentration of hydrogen sulfide in the gas resources [1,7]. Sulfolane is also utilized as a solvent for batteries, a plasticizer, and a solvent in the polymer industry, with an estimated annual production of 18,000 to 36,000 tons, reflecting its variability of function in different fields [1, 8].

Despite efforts to recover and reuse sulfolane in industrial processes, a small portion of the solvent is lost during each use cycle. For instance, the loss is estimated to be around five parts per million (ppm) in the UOP Sulfolane™ commercial process for extracting aromatics from a mixture of hydrocarbons [9]. This small fraction of sulfolane becomes significant when considering the huge amounts of the solvent used annually. In addition, sulfolane can be released into the environment through direct spillage or improper waste management practices, and reports of such incidents are available from different countries, including Canada and the US [8].

Multiple articles and reports have examined the fate, remediation, and environmental impacts of sulfolane [8,10–13]. In brief, due to its chemical stability, sulfolane persists in the environment for a long time when released, and its high solubility in water can lead to the contamination of groundwater resources. Furthermore, sulfolane's low adsorption on most soil types means that it can spread over large areas through groundwater after being released into the environment [10]. For example, at a contaminated site in Waterton, Canada, it is estimated that one square mile around the contamination source has been impacted [11].

Numerous studies have indicated that sulfolane exhibits toxicity in rats and mice. However, it is imperative to note that further research is necessary to substantiate these effects in the context of human exposure [14]. The mobility and toxicity of sulfolane have raised concerns over its spread into environmental media, and natural aerobic biodegradation is the major way for sulfolane to be removed when released into the environment.

However, the low oxygen levels in groundwater resources make biodegradation a slow process [15].

To assess the extent and levels of contamination, a sensitive, robust, and accurate analytical method is required. This method must also be versatile and applicable to different environmental matrices, and, if possible, it should not negatively impact the environment. Upon an examination of the literature, only one review article on the subject was identified, dating back more than 20 years [16]. Subsequently, numerous advancements in analytical techniques and methodologies have emerged and been applied in sulfolane analysis. This article aims to offer insights into the present state of sulfolane analysis and its progressive developments over the years and critically evaluate the methodologies employed.

To obtain the relevant literature on sulfolane analysis, the Web of Science database was utilized with appropriate keywords. A total of 49 articles were selected, which provided methods for analyzing sulfolane. Moreover, an exhaustive search for official methods and governmental guidelines for sulfolane analysis was conducted, which yielded two official methods of analysis. The reported parameters of the analytical methods were organized into different stages of the analytical process and are summarized in the subsequent sections.

2. Sampling of Environmental Matrices

Sampling in analytical chemistry is defined as the process of collecting a representative fraction of a material so that it can be analyzed to determine certain characteristics of the bulk of that material. This step is crucially important in the analytical process because of its immense and immediate effect on the validity of the results [17]. To achieve representativeness, different matrices require different sampling techniques. There are well-established official sampling procedures for semi-volatile compounds such as sulfolane in different matrices published by the US EPA which must be followed to ensure minimizing the error introduced into the analysis by inappropriate sampling [18,19].

3. Sample Preparation

Sample preparation is a critical step in sulfolane analysis, as it involves addressing the challenges of low analyte concentrations and potential co-contaminants that may be present in the sample. Similar to other analytical procedures, sample preparation plays a pivotal role in ensuring accurate and reliable results. The key parameters that vary among different sample preparation procedures include the extraction method, the inclusion of any concentration step, and the utilization of internal standards for error correction. These parameters for the reviewed publications are summarized in Table 2. In the subsequent sections, a selection of these papers is discussed which reveals useful insights into different types of extraction, their challenges, and how to deal with them.

Table 2. Summary of the sample preparation methods of the reviewed articles.

Sample Type	Method of Extraction	Extraction Phase	NO. of Extraction Aliquotes	Solvent-to-Sample Ratio (mL:mL or mL:g)	Centrifugation	Filtration	Salting out Agent (Saturation %)	Water Removal of Extract	Solvent Evaporation (Concentration Factor)	pH Adjustment	Internal Standard	Recovery	Source
Aqueous solution	AQ [1] direct injection	N/A	N/A	N/A	Yes	By filter paper	No	No	No	No	No	N/A	[20]
Aqueous solution	LLE [2]	DCM [4]	3	2:5	No	No	No	No	No	No	No	60 to 70	[21]
Aqueous solution	LLE [2]	DCM [4]	1	2:5	No	Yes	NaCl (80%)	No	No	No	Ethylene glycol butyl ether	N/A	[22]
Aqueous solution	LLE [2]	DCM [4]	1	1:1	No	No	No	No	No	No	No	80	[23]
Aqueous solution	LLE [2]	DCM [4]	1	2:5	No	N/A	No	No	No	No	No	N/A	[24]
Aqueous solution	LLE [2]	DCM [4]	1	2:1	No	No	No	No	No	No	No	80	[25]
Aqueous solution	LLE [2]	Ethyl acetate	1	2:1	No	No	No	No	No	No	No	N/A	[26]
Aqueous solution	LLE [2]	DCM [4]	1	3:5	No	Yes	No	No	No	No	No	80 ± 5	[27]
Aqueous solution	LLE [2]	DCM [4]	1	3:5	No	Yes	No	No	No	No	No	80 ± 5	[28]
Aqueous solution	LLE [2]	Toluene	1	2:1	No	No	No	No	No	No	Sulfolane-d8	N/A	[12]
Aqueous solution	LLE [2]	DCM [4]	1	3:5	No	Yes	No	No	No	No	No	80 ± 5	[29]
Aqueous solution	LLE [2]	DCM [4]	1	3:5	No	Yes	No	No	No	No	No	80 ± 5	[30]
Aqueous solution	LLE [2]	DCM [4]	1	3:5	No	Yes	No	No	No	No	No	N/A	[31]
Aqueous solution	LLE [2]	DCM [4]	1	3:5	No	Yes	No	No	No	No	No	80 ± 5	[32]
Aqueous solution	LLE [2]	DCM [4]	1	1:1	No	By 0.2 μm PTFE filter	No	No	No	No	No	N/A	[33]
Aqueous solution	LLE [2]	DCM [4]	1	1:1	No	By 0.2 μm PTFE filter	No	No	No	No	No	N/A	[34]
Aqueous suspension	AQ [1] direct injection	N/A	N/A	N/A	Yes	By 0.22 μm nylon filter	No	No	No	No	No	N/A	[35]

Table 2. *Cont.*

Sample Type	Method of Extraction	Extraction Phase	NO. of Extraction Aliquotes	Solvent-to-Sample Ratio (mL:mL or mL:g)	Centrifugation	Filtration	Salting out Agent (Saturation %)	Water Removal of Extract	Solvent Evaporation (Concentration Factor)	pH Adjustment	Internal Standard	Recovery	Source
Aqueous suspension	LLE [2]	DCM [4]	3	N/A	N/A	N/A	No	No	No	No	Sulfolane-d8 and nitrobenzene-d8	N/A	[36]
Aqueous suspension	LLE [2]	DCM [4]	1	1:10	No	By 0.22 µm Teflon filter paper	NaCl (100%)	No	No	No	No	N/A	[37]
Aqueous suspension	LLE [2]	DCM [4]	1	3:5	No	By 0.45 µm membrane syringe	No	No	No	No	No	N/A	[31]
Biochar	Soxhlet extraction	Ethyl acetate	6	N/A	No	No	No	No	N/A	No	N-butanol	43 to 50	[38]
Ethanol/water (1:1) mix	LLE [2]	Chloroform	1	1:5	No	No	No	By Na_2SO_4	No	No	Dicyclohexylmethanol	45.5	[39]
Homogenized water plant tissue mixture	LLE [2]	Toluene	3	5:1	Yes	By 0.2 µm cellulose acetate membrane filter	No	No	Yes (15)	No	No	80 ± 12	[40]
Homogenized water plant tissue mixture	LLE [2]	Toluene	3	5:1	Yes	By 0.2 µm cellulose acetate membrane filter	No	No	Yes (15)	No	No	80 ± 12	[41]
Homogenized water plant tissue mixture	LLE [2]	DCM [4]	3	2:5	No	By 0.2 membrane filter	No	No	No	No	No	50 to 60	[21]
Plasma	LLE [2]	Ethyl acetate	1	5:1	Yes	No	Yes	No	Yes (8)	by NaOH	Sulfolane-d8	93	[42]
Plasma	LLE [2]	Ethyl acetate	1	5:1	Yes	No	Yes	No	Yes (8)	by NaOH	Sulfolane-d8	74.4 to 88.7	[43]
Plasma	LLE [2]	Ethyl acetate	1	5:1	Yes	No	Yes	No	No	by NaOH	Sulfolane-d8	>85	[44]
PPSF polymer	SLE [3]	Acetonitrile	1	20:1	No	No	No	No	No	No	Dicyclohexylmethanol	N/A	[39]

Table 2. *Cont.*

Sample Type	Method of Extraction	Extraction Phase	NO. of Extraction Aliquotes	Solvent-to-Sample Ratio (mL:mL or mL:g)	Centrifugation	Filtration	Salting out Agent (Saturation %)	Water Removal of Extract	Solvent Evaporation (Concentration Factor)	pH Adjustment	Internal Standard	Recovery	Source
Soil	SLE [3]	Water	2	5:1	Yes	No	No	No	No	No	No	65 to 102	[45]
Soil	SLE [3]	Water	1 to 3	1:1 to 3:1	Yes	0.45 μm filter	No	No	No	No	No	82 to 99	[46]
Soil	SLE [3]	DCM [4]	1	1:1	Yes	No	No	By Na_2SO_4 and NaCl	No	No	Sulfolane-d8	80–120	[47]
Soil	SLE [3]	DCM [4]	3	5:2	Yes	Yes	No	By Na_2SO_4	If required	No	Sulfolane-d8	70–120	[48]
Soil	Soxhlet extraction	DCM [4]	1	30:1	No	No	No	By Na_2SO_4	Yes (3)	No	Sulfolane-d8	70–120	[48]
Soil slurry	AQ [1] direct injection	Water	N/A	N/A	Yes	No	No	No	No	No	No	N/A	[7]
Soil slurry/soil water mix	LLE [2]	DCM [4]	3	1:2	Yes	No	NaCl (80%)	By Na_2SO_4	Yes (N/A)	No	Dibenzothiophene	N/A	[49]
Soil water mix	LLE [2]	Toluene	3	5:1	Yes	By 0.2 μm cellulose acetate membrane filter	No	No	Yes (15)	No	No	126	[41]
Water	AQ [1] direct injection	N/A	0	N/A	Yes	No	No	No	No	No	No	N/A	[50]
Water	AQ [1] direct injection	Water	N/A	N/A	Yes	No	No	No	No	No	No	N/A	[51]
Water	LLE [2]	DCM [4]	1	1:1	Yes	By 0.2 membrane filter	No	By Na_2SO_4	No	No	No	N/A	[10]
Water	LLE [2]	DCM [4]	1	1:1	Yes	By 0.2 membrane filter	No	No	No	No	No	N/A	[52]
Water	LLE [2]	DCM [4]	3	1:20	No	By glass wool	No	By Na_2SO_4	Yes (15)	By NaOH to pH 10	No	N/A	[53]
Water	LLE [2]	Toluene	3	5:1	Yes	By 0.2 μm cellulose acetate membrane filter	No	No	Yes (15)	No	No	127	[41]

Table 2. *Cont.*

Sample Type	Method of Extraction	Extraction Phase	NO. of Extraction Aliquotes	Solvent-to-Sample Ratio (mL:mL or mL:g)	Centrifugation	Filtration	Salting out Agent (Saturation %)	Water Removal of Extract	Solvent Evaporation (Concentration Factor)	pH Adjustment	Internal Standard	Recovery	Source
Water	LLE [2]	DCM [4]	1	1:1	No	By 0.2 membrane filter	No	By Na_2SO_4	No	No	No	N/A	[54]
Water	LLE [2]	DCM [4]	1	1:10	No	No	NaCl (100%)	No	No	By Na_2CO_3 to pH 5–8	No	N/A	[55]
Water	LLE [2]	DCM [4]	1	1:10	No	No	NaCl (100%)	No	No	By Na_2CO_3 to pH 5–8	No	N/A	[56]
Water	LLE [2]	DCM [4]	1	5:2	No	By 0.45 μm PTFE	No	No	No	No	No	N/A	[15]
Water	LLE [1]	1,2-dichloroethane	1	1:1	No	No	No	No	No	No	No	N/A	[57]
Water	LLE [2]	DCM [4]	2	1:1	No	No	Yes	By Na_2SO_4	Yes (200)	Yes (pH < 2)	Sulfolane-d8	80–120	[47]

[1] Aqueous extract. [2] Liquid–liquid extraction. [3] Solid–liquid extraction. [4] Dichloromethane. N/A—not applicable.

3.1. Extraction

Extraction is a widely used process that takes advantage of the differences in chemical distribution between two immiscible phases to separate the constituents of a sample. The main objective of extraction is to partition the analytes of interest into one phase while keeping other sample constituents in the other phase, resulting in a simplified sample composition and reduced interference from other compounds. The extraction itself might lead to analyte concentration if the volume of the extractant is lower than the sample volume and extraction is exhaustive, i.e., most of the analyte is removed from the sample into the extraction phase [17].

For sulfolane analysis, two common extraction techniques are liquid–liquid extraction (LLE) and solid–liquid extraction (SLE). In both techniques, a solvent is added to the sample phase, and the analyte partitions into the solvent, which is the extraction phase. The recovery of the extraction process is defined as the fraction of the original analyte that is removed from the sample phase [17]. The goals of the extraction process are to isolate and concentrate the analyte and to exchange the sample matrix for one that is compatible with the analytical instrument.

There are several factors that can affect recovery in sulfolane extraction, which will be discussed in the following sections. Understanding these factors is crucial in optimizing the extraction process for sulfolane analysis, ensuring accurate and reliable results.

3.1.1. Extraction Solvent

The choice of extraction solvent plays a crucial role in the recovery and sensitivity of the analysis in various ways. The partitioning of the analyte into the extraction phase is influenced by the characteristics and volume of the extraction solvent, which, in turn, impacts the fraction of the analyte that gets extracted. To increase the extraction efficiency, multiple aliquots of the solvent can be used to extract the sample multiple times. This will result in a higher recovery of the analyte, and by removing most of the analyte into the extraction phase, exhaustive extraction can be achieved. Additionally, higher sensitivities can be achieved by increasing the amount of analyte extracted.

Blowing down the solvent is an effective way to increase the concentration of the extract and improve sensitivity, especially when combined with multiple aliquot extractions that maximize the amount of extracted analyte. However, this approach may not be practical with high-boiling-point solvents such as water because of the longer evaporation time and higher energy requirements. More importantly, when the boiling point of the solvent is high, there is a greater risk of analyte loss due to evaporation, particularly if the boiling point of the analyte is lower than that of the solvent. On the other hand, solvents with low boiling points can be easily evaporated to concentrate the extract or dried to reconstitute the sample in another solvent, resulting in a higher concentration and a higher sensitivity.

It is important to note that the choice of the extraction solvent also affects gas chromatography (GC) analysis. Typically, the extraction solvent is the same as the final sample solvent. If high-boiling-point solvents are used for the extraction, it can result in co-elutions with some sample components, leading to the inability to detect early eluting analytes. This highlights the need to carefully consider the boiling point and other characteristics of the extraction solvent in order to optimize the recovery and sensitivity of the analysis. The specific solvents used for the extraction of sulfolane are discussed in the following sections.

Dichloromethane

As summarized in Table 2, Dichloromethane (DCM) has been commonly used for the extraction of sulfolane from solid and liquid samples in numerous studies. Despite an extensive literature research, no information on the solubility of sulfolane in DCM was found. Nevertheless, reported recovery rates for sulfolane extraction using DCM ranged from 50% to 80%, which falls within the acceptable range based on the only two available official methods for sulfolane determination in environmental samples [47,48]. Despite the acceptable extraction performance demonstrated in various studies, DCM is widely recognized as one of the most harmful solvents to the environment [58]. One of the reasons for this negative reputation is that DCM has been found to contribute to the delay in the recovery of the ozone layer in the Earth's stratosphere [59].

For water samples, most studies utilized a single aliquot of DCM for extraction, with DCM-to-sample ratios ranging from 1:20 [53] to 5:2 [15], with most being around 1:1. Notably, except for Kim et al., none of the studies exploited the easy evaporation capability of DCM to further concentrate the final samples [53]. The decision to exclude this step could be attributed to achieving adequate sensitivity in their analyses. However, if a single-step extraction is intended, using a lower ratio of extraction solvent to sample could result in a higher concentration of the extract. Despite extracting a lower fraction of analytes, this approach can lead to higher sensitivities. Moreover, by utilizing more efficient extraction procedures, similar sensitivities can be achieved using less solvent, thereby reducing the negative environmental impact of the analysis.

On the other hand, in a study conducted by Doucette et al., water samples were extracted with three aliquots of DCM at a DCM-to-sample ratio of 2:5, and the combined extract was reserved for analysis [21]. The reported recovery ranged from 60% to 70%, which barely meets the acceptance criteria set by the only two official guidelines for the water sample analysis of sulfolane [47,48]. The use of three aliquots of solvent for extraction should result in a higher extraction efficiency. However, when the recovery reported by Doucette et al. [21] is compared to similar studies conducted using single-aliquot extraction with lower cumulative solvent volumes, such as the studies by Izadifard et al. [23] and Khan et al. [28], the achieved recovery is 10% to 20% lower. This is unexpected, as the absolute recovery should be directly proportional to the fraction of the analyte partitioned from the sample into the extraction phase, and, theoretically, more aliquots of extraction and higher cumulative volumes should result in higher recoveries.

Since the samples used in the study by Doucette et al. [21] were water samples that did not contain particles for sulfolane to adsorb onto, and the concentrations of sulfolane were not high enough to saturate DCM and decrease sensitivity, one explanation for the lower recovery could be the possible difference in extraction temperature, as can be inferred from the study by Li et al. [57]. In this study, they aimed to inspect the suitability of 1,2-dichloroethane for the extraction of sulfolane from an aqueous phase, and they created a ternary system consisting of water, 1,2-dichloroethane, and sulfolane in varying proportions. They measured the distribution of sulfolane between the organic and aqueous phases and found that at low concentrations of sulfolane, sulfolane tends to be enriched in the organic phase up to 120 times. Additionally, there is an inverse relationship between sulfolane enrichment in the organic phase and temperature in the range of 15 °C to 35 °C. The effect is significant, as the enrichment decreased from 120:1 at 15 °C to 32:1 at 35 °C [57]. Although there is no such study performed for the DCM extraction of sulfolane, it can be expected that DCM and 1,2-dichloroethane behave similarly due to their similar structure. This relationship between temperature and sulfolane distribution could be the reason for the significant difference in recoveries observed when using seemingly similar extraction procedures. Furthermore, it would be interesting to explore if and to what extent the distribution of sulfolane in other water-immiscible solvents, such as toluene and ethyl acetate, is affected by temperature.

In another study conducted by Greene et al., DCM was used for soil sample extraction [7]. The extraction was carried out using a Soxhlet apparatus for 6 h. Soxhlet extraction is an example of a solid–liquid extraction technique where the analyte is continuously removed by the solvent at its boiling point. This method allows for higher recoveries, especially for challenging samples, due to the higher temperature and circulation of distilled solvent over the sample [60]. Following the extraction, the solvent was concentrated using a rotary evaporator, as described by Greene et al. [51]. Notably, this extraction method is similar to the official soil extraction method used by the government of Alaska, which will be discussed later in this section.

DCM is also the solvent recommended by the only two official guidelines available for sulfolane analysis. In the lab manual for sulfolane analysis by the government of British Columbia, a 100 mL water sample is adjusted to a pH below 2 and then extracted with two aliquots of 100 mL of DCM. The extracted samples are then treated with sodium sulfate to remove water, followed by the complete evaporation of the solvent and reconstitution of the sample in 1 mL of isooctane [47]. Although this procedure results in a concentration of sulfolane in the sample by around a factor of 200, the environmental impact of using 200 mL of DCM for every sample is significant. Scaling down the extraction procedure and reevaluating the need for such a high analyte concentration rate may be strategies to reduce solvent usage, especially considering that reasonably low limits of detection have been achieved with simpler and less environmentally hazardous procedures [41]. However, if these methods are to be replaced, extensive validation procedures must be undertaken to ensure their reliability. In contrast, the government of Alaska has not specified any particular method of extraction for water samples but has designated the required performance parameters for the extraction [48].

For soil samples, the extraction procedure described in the BC lab manual involves adding 1 mL of water to 10 to 12 g of a field-moist sample, followed by extraction with 10 mL of DCM. Sodium sulfate and sodium chloride are added to remove water, and after centrifugation, 1 mL of the extraction phase is reserved for analysis [47]. Although this method uses much less DCM compared to the official water method, it does not include a concentration step. It may be possible to increase sensitivity by taking a larger portion of the DCM extract and concentrating it through evaporation.

On the other hand, the government of Alaska suggests using EPA method 3550C or, preferably, 3540C for soil samples [48]. EPA method 3550C involves mixing 40 g of a soil sample with 60 g of sodium sulfate to remove water, followed by three 100 mL aliquots of DCM extraction while being sonicated for 3 min. The extracts are then filtered and centrifuged to remove any particles [61]. EPA method 3540C for soil samples involves mixing 10 g of soil sample with sodium sulfate to remove water, followed by Soxhlet extraction with 300 mL of DCM for 16 to 24 h. The solvent is then evaporated and reconstituted in 100 mL to 125 mL of DCM [62]. Both methods use larger amounts of solvents and achieve a subtle increase in the concentration of the analyte in the final extracts. Furthermore, Soxhlet extraction is time-consuming and energy-intensive, and it is prone to analyte loss for volatile co-contaminants.

For plant tissues, the Alaskan government requires the sample to be ground, frozen using liquid nitrogen, pulverized, and then extracted with water, followed by extraction with DCM [48]. This method is based on the method by Headley et al., discussed later in Section "Toluene", with the difference being that toluene has been replaced by DCM as the extraction phase [41].

Water

In order to optimize water extraction conditions for soil samples, Brandao et al. investigated the effects of extraction time, number of extraction steps, and soil-to-water ratio on recovery [46]. They concluded that increasing the extraction time from 30 to 90 min did not significantly affect the recovery. However, increasing the water-to-soil ratio from 1:1 to 3:1 resulted in an increase in recovery from 82% to 93%. Additionally, three cycles of extraction with a water-to-soil ratio of 2:1 resulted in increased recovery from 85.2% for the first cycle to 97.2% for the second cycle, and 98.9% for the third cycle. To confirm the remaining sulfolane content in the soil after water extraction, Soxhlet extraction with DCM was used [46]. Although using two or three aliquots of water for the extraction resulted in better recoveries, it may negatively impact sensitivity due to the impracticality of evaporating water and concentrating the final extract. In cases where higher sensitivities are required, a single aliquot of water can be used. However, using only one aliquot of water may result in incomplete extraction, as the recovery is dependent on the partitioning equilibrium, which is influenced by the distribution constant [60]. The distribution constant can be affected by various factors, such as temperature, the composition of the soil sample, the presence of ionic species in the soil sample, and co-contaminants, all of which add uncontrollable parameters to the analysis.

An alternative solution to address this issue is to use a water-immiscible solvent to extract sulfolane from the aqueous extract, as demonstrated by Fedorak, 1996 [49]. While this approach may be beneficial in certain scenarios, such as when there are numerous hydrophobic compounds that require clean-up, the limitation of water as an extraction phase in terms of its inability to extract hydrophobic co-contaminants remains unresolved. Moreover, the addition of an additional extraction step introduces another potential source of errors, which can adversely impact analytical reliability.

Water samples can be directly injected into analytical instruments for analysis without any pretreatment, if the analyte concentration is high enough for detection. This approach, known as direct injection of aqueous samples, eliminates the sample extraction step, which minimizes errors introduced during extraction. However, the samples must be clean and free from excessive constituents, as this method is susceptible to contamination of the analytical instrument. Therefore, this method may not be suitable for dirty samples such as wastewater samples. To extend the applicability of this technique, centrifugation and sample filtration can be employed to separate the aqueous phase for samples that may contain particles, such as soil slurries [7]. In cases where filtration is used, it is important to confirm that the analyte does not adsorb onto the filter, as demonstrated by Yu et al. in their study for nylon filters [35].

In an uncommon and novel extraction method, water was used in an ordinary espresso machine to extract sulfolane from biochar. The authors concluded that ground biochar extraction with water in an espresso machine is more efficient than ethyl acetate extraction with Soxhlet for six solvent cycles or dispersion of biochar in ethyl acetate for 48 h with initial ultrasonication for 45 min. They also reported a loss of sulfolane when they tried to evaporate water at 60 °C or 104 °C in the presence of biochar due to unknown reasons [38]. This method of extraction is a simple and inexpensive setup for pressurized liquid extraction. Considering the 10 to 20 bars of pressure offered by these machines, they can be considered low-pressure liquid extractors. However, since the extraction is performed with water just below its boiling point, it cannot be used for the simultaneous extraction of volatile co-contaminants. Additionally, although the authors achieved reproducible recoveries based on their limited set of experiments, it is interesting to see how reproducibly this machine performs over time, as most of these machines lose pressure as they age.

Toluene

Toluene has been utilized as a solvent for extracting sulfolane from water samples [41]. Notably, it has been employed for the back-extraction of sulfolane from soil or plant tissue extracts. The reported recoveries for direct water extraction and the extraction of water extract from soil samples were remarkably high at 127% and 126%, respectively [41]. However, no explanation was provided for these high recoveries, but it is possible that the non-symmetric solubility of the toluene–water system formed during extraction [63] and the high extraction efficiency of toluene for sulfolane contributed to these results. It should be noted that the high boiling point of toluene may hinder the analysis of volatile co-contaminants by GC.

The high extraction efficiency of toluene for sulfolane was effectively utilized in the extraction of sulfolane from plant tissue samples by Headley et al. [41]. In their procedure, they homogenized plant tissue in Mili-Q water to extract the sulfolane into the aqueous phase. In a slightly different approach, they ground the plant tissue to a fine powder using liquid nitrogen and then suspended it in Mili-Q water. In both cases, the tissue was left in water for 40 min with intermittent swirling. Subsequently, they centrifuged and filtered the supernatant to remove any particles. Next, they back-extracted the aqueous extract with three aliquots of toluene. The toluene portions were combined, and the extract was concentrated 15 times using nitrogen blowdown before being reserved for analysis by GC-MS. The average recovery achieved was 80.3% with an RSD of 11%. Remarkably, the method's performance was maintained over a period of 18 months, indicating its ruggedness [41].

The decision to use water as an intermediate extraction solvent for plant tissue is reasonable due to the hydrophilic nature of plant tissue. By leaving the tissue in an aqueous solution, water can penetrate and swell the tissue, allowing for better dissociation and partitioning of the analytes into the bulk of water. Aprotic organic solvents, on the other hand, do not swell the tissue and poorly penetrate it, resulting in a poor extraction efficiency. Moreover, plant tissue may contain hydrophobic compounds that can interfere with the analysis, and using water as an intermediate step, with its low solubility for these compounds, is a preferable choice over other protic solvents such as methanol. Additionally, the use of water as an intermediate solvent also serves as a clean-up step. However, the relatively lower recovery rate of 80% compared to soil and water samples in the study suggests possible adsorption of sulfolane onto the plant tissue, which needs to be monitored when developing a method for plant tissue analysis.

Ethyl Acetate

Ethyl acetate has been successfully used for plasma sample extraction with good recoveries, as demonstrated in a study by Versace et al. [42]. Considering that plasma is a complex aqueous solution with many constituents, it is reasonable to explore the use of ethyl acetate for sulfolane extraction from water samples, and, potentially, environmental samples. Moreover, ethyl acetate is known to have minimal impact on the environment [58].

In the study by Versace et al. [42], the extraction procedure involved adjusting the pH of 100 microliters of plasma by adding 50 microliters of 1 M sodium hydroxide, resulting in a highly basic pH that caused the aggregation of plasma proteins. Subsequently, the plasma was extracted with 500 microliters of ethyl acetate, followed by centrifugation to separate the organic phase. A total of 50 microliters of isopropanol were then added to the ethyl acetate extract, and the ethyl acetate was evaporated using N_2 gas to concentrate the extract approximately five times, while the remaining isopropanol extract was reserved for analysis. Despite the complexity of the plasma sample and the abundance of constituents in the matrix that could potentially adsorb sulfolane, the study achieved a remarkable recovery rate of 93%.

A notable feature of the described extraction procedure is the use of small volumes of samples and solvents, which is evident upon inspection. Additionally, as mentioned earlier, the environmental impact of ethyl acetate is comparatively low. When compared to the government of Alaska's official method for sulfolane extraction from soil samples, the solvent usage in the study by Versace et al. is three orders of magnitude lower [42,48]. Furthermore, similar studies, such as a study by Silinski et al., support the possibility of using ethyl acetate for sulfolane analysis [43]. In fact, Heydari et al. used ethyl acetate for water sample extraction, although no information about the extraction performance was provided in their publication [26]. However, it is worth noting that DCM was used as a replacement for ethyl acetate in their recent publications [13,30]. This may be due to the inadequate extraction performance of ethyl acetate in their procedure, possibly resulting from the lack of the addition of a salt or ionic strength modifier (discussed in Section 3.1.2) [26].

Other Solvents

Acetonitrile has been utilized as an extraction solvent to extract residual sulfolane content from ground polyphenyl sulfone polymers, as reported by Eckardt et al. [39]. In the same study, chloroform was also used for extracting sulfolane from an ethanol–water (50:50) mixture to analyze the sulfolane leaching extent from bottles using the ethanol–water mixture as a food simulant. Although efficiency parameters were not reported, all three solvents—ethanol, acetonitrile, and chloroform—have the potential to be used as extraction solvents for sulfolane [39]. It is worth noting that while chloroform is as hazardous to the environment as DCM, acetonitrile is less hazardous, and ethanol is considered one of the most environmentally friendly solvents [58].

On the other hand, 1,2-dichloroethane has been investigated as an extraction solvent for sulfolane from water samples. However, it poses environmental hazards similar to DCM and has a higher boiling point, resulting in longer blowout time, greater risk of analyte loss, and increased risk of co-elution with volatile compounds in GC analysis.

3.1.2. Dissolved Electrolyte Effect

In two-phase solvent systems used for extraction, a fraction of the organic phase can dissolve in the water phase. This fraction can be significant, reaching up to 8% for solvents like ethyl acetate [64], while for highly non-polar solvents like toluene, it is typically below 1% [63]. This dissolved organic phase can alter the distribution coefficient and reduce the partitioning of the solute from water into the organic phase. To mitigate this effect, an electrolyte can be added to the aqueous phase [64], which limits the fraction of the organic phase that dissolves in water and can improve the extraction efficiency.

For instance, sodium chloride at 80% of its saturation concentration [22,49] or at its saturation concentration [37,55,56] has been used to enhance the partitioning of sulfolane into DCM, which serves as the organic phase. Similarly, in the study conducted by Versace et al., adjusting the pH of water by adding sodium hydroxide increased the amount of electrolytes dissolved in the water, which consequently improved the extraction efficiency of ethyl acetate [42].

3.1.3. pH Adjustment

If a species dissociates at the pH of the extraction, only the undissociated portion of the analyte would take part in the partitioning equilibrium, and the ionized species resulting from dissociation would stay in the aqueous phase [17]. Hence, it is crucial to adjust the pH in a way that most of the analyte stays neutral for the extraction procedure. However, because of the high pK_a value (12.9), sulfolane does not dissociate significantly in the pH range typical of environmental samples. Hence, pH adjustment might not be necessary for most samples. It should be pointed out that it is required by the official method suggested by the government of BC for water samples [47]. One of the possible reasons for using pH adjustment in sulfolane analysis might be to limit the extraction of matrix concomitants. For instance, this principle has been exploited by Versace et al., Shipkowski et al. and Silinski et al. in plasma samples to aggregate and discard proteins present in the matrix [42–44].

4. Separation and Analysis Techniques

Despite the various extraction methods discussed in the literature, there has been limited diversity in the separation and analytical techniques utilized for the determination of sulfolane. Almost all researchers have employed GC to analyze sulfolane due to the low sensitivity of common LC detectors for this compound. Sulfolane does not absorb UV light at wavelengths that can be distinguished from the mobile phase, and, thus, cannot be detected by DAD detectors. Furthermore, it is not efficiently ionized by ESI, leading to the inadequate sensitivity of LC-MS systems [40]. In contrast, sulfolane is completely compatible with the two most widely used GC detectors: FID and MS. In the following sections, the conditions and configurations reported for the GC analysis of sulfolane are summarized.

4.1. GC Inlet

All of the reviewed studies utilized the split/splitless type of inlet. In this inlet system, the temperature must be high enough to causes all sample components to have non-insignificant vapor pressure so that they can be transferred effectively into the column. As shown in Table 3, temperatures in the range of 165 °C to 300 °C were successfully employed for sulfolane analysis.

Table 3. Summary of separation and determination methods of sulfolane.

Sample Type	Sample Solvent	Carrier Gas	Carrier Gas Flow Rate/Pressure	Mode of Injection	Injection Volume	Inlet Temperature (°C)	Type of Column	Column Stationary Phase	Column Dimensions	Detector	LLOD [4]	LLOQ [5]	Detector Temperature (°C)	Source
Aqueous solution	DCM [1]	N_2	10 mL/min	N/A	N/A	N/A	WCOT [2]	DB-5	30 m × 0.45 mm × 0.25 µm	FID	N/A	N/A	N/A	[21]
Aqueous solution	Water	N_2	360 mL/min	Splitless	N/A	280	SCOT [3]	OS-138	7.5 m × 0.51 mm	FID	6 mg/L	N/A	N/A	[20]
Aqueous solution	DCM [1]	He	1 mL/min	Splitless	1	250	WCOT [2]	DB-WAXETR	30 m × 0.32 mm × 1 µm	MS	N/A	N/A	N/A	[22]
Aqueous solution	DCM [1]	He	250 KPa	Splitless	2	165	WCOT [2]	ZB-5	N/A	FID	1 mg/L	N/A	250	[23]
Aqueous solution	DCM [1]	He	10 mL/min	Splitless	2	250	WCOT [2]	ZB-5	N/A	FID	N/A	N/A	320	[24]
Aqueous solution	DCM [1]	He	250 kPa	Splitless	2	165	WCOT [2]	ZB-5	N/A	FID	1 mg/L	N/A	250	[25]
Aqueous solution	Ethyl acetate	N/A	N/A	N/A	N/A	N/A	WCOT [2]	ZB-5	N/A	FID	N/A	N/A	N/A	[26]
Aqueous solution	DCM [1]	He	1.07 mL/min	Splitless	1	250	WCOT [2]	ZB-5	N/A	MS	10 µg/L	N/A	N/A	[27]
Aqueous solution	DCM [1]	He	1.07 mL/min	Splitless	1	250	WCOT [2]	ZB-5	N/A	MS	10 µg/L	N/A	N/A	[28]
Aqueous solution	Toluene	He	1 mL/min	Splitless	0.5	285	WCOT [2]	DB-5	30 m × 0.25 mm × 0.25 µm	MS	20 µg/L	70 µg/L	Ion source: 285	[12]
Aqueous solution	DCM [1]	He	1.07 mL/min	Splitless	1	250	WCOT [2]	ZB-5	N/A	MS	10 µg/L	N/A	N/A	[29]
Aqueous solution	DCM [1]	He	1.07 mL/min	Splitless	1	250	WCOT [2]	ZB-5	N/A	MS	10 µg/L	N/A	N/A	[30]
Aqueous solution	DCM [1]	He	N/A	Splitless	1	165	WCOT [2]	ZB-5	N/A	FID	0.3 mg/L	1 mg/L	330	[65]
Aqueous solution	DCM [1]	He	1.07 mL/min	Splitless	1	250	WCOT [2]	ZB-5	N/A	MS	10 µg/L	N/A	N/A	[32]
Aqueous solution	DCM [1]	He	1.07 mL/min	Splitless	1	165	WCOT [2]	ZB-5	N/A	FID	N/A	N/A	330	[33]
Aqueous solution	DCM [1]	He	1.07 mL/min	Splitless	1	165	WCOT [2]	ZB-5	N/A	FID	N/A	N/A	330	[34]
Aqueous suspension	Water	He	N/A	Splitless	1	165	WCOT [2]	ZB-5	N/A	FID	1 mg/L	N/A	250	[35]
Aqueous suspension	DCM [1]	N/A	N/A	Pulsed-splitless	N/A	N/A	WCOT [2]	RTX-200	30 m	MS	N/A	40 µg/L	N/A	[36]
Aqueous suspension	DCM [1]	N_2	N/A	Splitless	1	200	WCOT [2]	Stabliwax	30 m × 0.53 × 1 µm	FID	N/A	N/A	250	[37]

Table 3. *Cont.*

Sample Type	Sample Solvent	Carrier Gas	Carrier Gas Flow Rate/Pressure	Mode of Injection	Injection Volume	Inlet Temperature (°C)	Type of Column	Column Stationary Phase	Column Dimensions	Detector	LLOD [4]	LLOQ [5]	Detector Temperature (°C)	Source
Aqueous suspension	DCM [1]	He	N/A	Splitless	1	N/A	WCOT [2]	5-MSI	N/A	FID	N/A	N/A	330	[31]
Biochar	Ethyl acetate	He	N/A	Split (10:1)	1	N/A	WCOT [2]	ZB-Wax-Plus	30 m × 0.32 mm × 0.5 μm	FID	10 mg/L	N/A	N/A	[38]
Ethanol/water (1:1) solution	Chloroform	He	1 mL/min	Splitless	1	N/A	WCOT [2]	HP-5	30 m × 0.25 mm × 0.25 μm	MS	10 μg/Kg of Ethanol/water (1:1)	N/A	N/A	[39]
Homogenized water plant tissue mixture	Toluene	He	25 cm/s	Splitless	N/A	250	WCOT [2]	DB-5	25 m × 0.25 mm × 0.25 μm	MS	90 ng per gram of wet plant tissue	300 ng per g of wet plant tissue	Ion source: 280	[40]
Homogenized water plant tissue mixture	Toluene	He	25 cm/s	Splitless	N/A	250	WCOT [2]	DB-5	25 m × 0.25 mm × 0.25 μm	MS	90 ng per gram of wet plant tissue	300 ng per g of wet plant tissue	Ion source: 280	[41]
Homogenized water plant tissue mixture	DCM [1]	N_2	10 mL/min	N/A	N/A	N/A	WCOT [2]	DB-5	30 m × 0.45 mm × 0.25 μm	FID	N/A	N/A	N/A	[21]
Plasma	Isopropanol	He	1 mL/min	Splitless	1	250	WCOT [2]	ZB-5	15 m × 0.25 mm × 0.25 μm	MS/MS	N/A	20 μg/L	Ion source: 200	[42]
Plasma	Isopropanol	He	1 mL/min	Split (50:1)	1	250	WCOT [2]	DB-5	30 m × 0.25 mm × 0.25 μm	MS	0.516 μg/L	20 μg/L	Ion source: 230	[43]
Plasma	Ethyl acetate	He	1 mL/min	Split (50:1)	1	250	WCOT [2]	DB-5	30 m × 0.25 mm × 0.25 μm	MS	1.25 μg/L	N/A	Ion source: 230	[44]
PPSF polymer	Acetonitrile	He	1 mL/min	Splitless	1	N/A	WCOT [2]	HP-5	30 m × 0.25 mm × 0.25 μm	MS	N/A	N/A	N/A	[39]
Soil	Water	He	25 mL/min	Splitless	2	250	Packed	Tenax-GC coated with 5% polyphenyl ether	1.2 m × 0.32 cm	FID	1 mg/L	N/A	250	[45]
Soil	DCM [1]	He	250 KPa	Splitless	1	250	WCOT [2]	ZB-5	N/A	MS	<1 mg/L	<1 mg/L	330	[46]

Table 3. *Cont.*

Sample Type	Sample Solvent	Carrier Gas	Carrier Gas Flow Rate/Pressure	Mode of Injection	Injection Volume	Inlet Temperature (°C)	Type of Column	Column Stationary Phase	Column Dimensions	Detector	LLOD [4]	LLOQ [5]	Detector Temperature (°C)	Source
Soil slurry	Water	He	24 mL/min	Splitless	2	250	Packed	Tenax-GC coated with 5% polyphenyl ether	2 m × 0.3 cm	FID	0.5 mg/L	N/A	250	[7]
Soil slurry or soil water mixture	DCM [1]	N/A	N/A	Split (20:1)	N/A	250	WCOT [2]	DB-5	25 m × 0.25 mm × 0.25 µm	FID	N/A	N/A	250	[49]
Soil water mixture	Toluene	He	25 cm/s	Splitless	N/A	250	WCOT [2]	DB-5	25 m × 0.25 mm × 0.25 µm	MS	90 ng per gram of wet plant tissue	N/A	Ion source: 280	[41]
Water	Water	N_2	30 mL/min	Splitless	2	250	Packed	Tenax-GC coated with 5% polyphenyl ether	1.8 m × 0.32 cm	FID	5 mg/L	N/A	250	[50]
Water	Water	He	25 mL/min	Splitless	2	250	Packed	Tenax-GC coated with 5% polyphenyl ether	1.2 m × 0.32 cm	FID	1 mg/L	N/A	250	[51]
Water	DCM [1]	He	1.7 mL/min	N/A	N/A	300	WCOT [2]	DB-5	30 m × 0.25 mm × 0.25 µm	FID	N/A	N/A	350	[10]
Water	DCM [1]	He	N/A	N/A	N/A	N/A	WCOT [2]	DB-5	30 m × 0.25 mm × 0.25 µm	FID	N/A	N/A	N/A	[52]
Water	DCM [1]	He	1.7 mL/min	Split (N/A)	5	300	WCOT [2]	DB-5	30 m × 0.25 mm × 0.25 µm	FID	N/A	N/A	350	[53]
Water	Toluene	He	25 cm/s	Splitless	N/A	250	WCOT [2]	DB-5	25 m × 0.25 mm × 0.25 µm	MS	1 ng/mL	N/A	Ion source: 280	[41]
Water	DCM [1]	He	1.7 mL/min	N/A	N/A	300	WCOT [2]	DB-5	30 m × 0.25 mm × 0.25 µm	FID	N/A	N/A	350	[54]
Water	DCM [1]	He	N/A	N/A	0.2	200	WCOT	Stabliwax	30 m × 0.53 × 1 µm	FID	N/A	N/A	250	[55]
Water	DCM [1]	He	N/A	N/A	0.2	200	WCOT [2]	Stabliwax	30 m × 0.53 × 1 µm	FID	N/A	0.44 mg/L	250	[56]
Water	DCM [1]	N/A	N/A	N/A	N/A	N/A	N/A	N/A	N/A	FID	1 mg/L	N/A	N/A	[15]
Water	1,2-dichloroethane	N_2	4.5 mL/min	Split (8:1)	0.2	250	WCOT [2]	DB-FFAP	30 m × 0.53 × 1 µm	FID	N/A	N/A	250	[57]

[1] Dichloromethane. [2] Wall-coated open tubular. [3] Support-coated open tubular. [4] Lower limit of detection. [5] Lower limit of quantification. N/A—not applicable.

With the split/splitless inlet system, different modes of injection can be used for introducing the sample into the GC column. Several studies have utilized split mode injection, as shown in Table 3. In this injection mode, the sample is vaporized, but most of the vapor exits in the inlet, and only a small fraction of the vapor is transferred into the capillary column. This injection mode is ideal for injecting high-concentration samples, often eliminating the need for sample dilution. Additionally, owing to a narrow injection band, narrower peaks can be achieved in this mode [66]. However, since the concentration of sulfolane is typically low in environmental samples, more concentration must typically be performed in the sample preparation step to analyze the samples in this mode.

In most studies, the splitless mode of injection has been used, as indicated in Table 3. In this mode, most of the sample vapor is carried by the carrier gas into the column. Since the transfer of the vapor into the column could take several seconds to minutes depending on the carrier gas flow and inlet volume, the initial temperature of the column oven must be lower than the boiling point of the sample solvent to refocus the injection band on the column head through the solvent's effect and prevent peak widening due to wide injection [66]. This type of injection results in higher sensitivity since more analyte is transferred into the instrument and is useful when the analyte concentration is low in the sample.

In a study by [36], a pulsed splitless mode of injection was utilized. In this mode, the pressure and, consequently, the carrier gas flow rate are raised temporarily. The vaporized sample will transfer into the column faster and more effectively by the raised flow rates, and the formed band will be narrower. Additionally, since the pressure of the inlet is higher at the time of injection, the solvent expansion volume will be lower, and higher volumes of sample can be injected. Both the higher injection volume and narrower band could increase sensitivity [67].

4.2. GC Column

In earlier publications, there are reports of using packed columns for GC analysis of sulfolane. However, recent studies have exclusively employed capillary columns due to their higher separation efficiency. In the following sections, the properties of these capillary columns are outlined. For complete specifications of the columns used in each study, please refer to Table 3.

4.2.1. Stationary Phase

The majority of studies on sulfolane have utilized non-polar columns coated with (5%-phenyl)-methylpolysiloxane as the stationary phase. This type of stationary phase results in a weak retention of sulfolane, and, consequently, the elution time is not long. However, if other polar compounds or volatile non-polar compounds are present in the sample, they may co-elute with sulfolane. On the other hand, several researchers have employed columns with more polar stationary phases, specifically polar polyethylene glycol [22,38] and mid-polar trifluoropropyl methyl polysiloxane [36].

Due to the high polarity of sulfolane, polyethylene glycol columns retain it strongly, and the analysis time is typically longer due to the lower maximum temperature limit of these columns. Therefore, the use of polyethylene glycol columns is only justified when there are polar components in the sample that need to be separated from sulfolane for reliable analysis.

In the study by [36], mid-polar trifluoropropyl methyl polysiloxane columns were used to exclude hydrocarbon co-contaminants that might have been present in the samples.

4.2.2. Column Dimensions

The dimensions of the columns used for sulfolane analysis vary in length from 15 m to 30 m, in inner diameter (ID) from 0.25 mm to 0.53 mm, and in film thickness from 0.25 µm to 1 µm (Table 3). While longer columns may provide a better separation, they result in a prolonged analysis time, which can lead to band broadening and reduced sensitivity.

Therefore, it is recommended to choose the shortest column that can separate the sample components [66].

Columns with narrower diameters and thinner films result in a higher chromatographic resolution but have a lower sample capacity. This is not a major issue for detection with MS, which is characterized by a high sensitivity. In the case of FID detectors, thicker stationary phases may be preferred to achieve a higher sensitivity through the use of larger injection volumes [66].

4.3. GC Detectors

As previously mentioned, both MS and flame ionization detectors are commonly used for sulfolane analysis. However, MS provides superior sensitivity and selectivity compared to FID. In fact, reported limits of detection (LODs) for MS range from 0.516 μg/L to 20 μg/L, while FID LODs range from 500 μg/L to 6000 μg/L in aqueous samples. Additionally, MS allows for the use of isotopically labeled internal standards, which results in more reliable and reproducible results.

Electron impact (EI) was the preferred method for all mass spectrometric techniques. The mass spectrum shown in Figure 2 indicates that the molecular ion for sulfolane appears at 120 m/z, with two abundant fragment ions at m/z 41 and 56, corresponding to the loss of SO_2CH_3 and SO_2, respectively. Different publications have used either the ion at m/z 120 or 41 for quantification, but Headley et al. reported five times higher sensitivity when using m/z 41 as the quantifier ion [40].

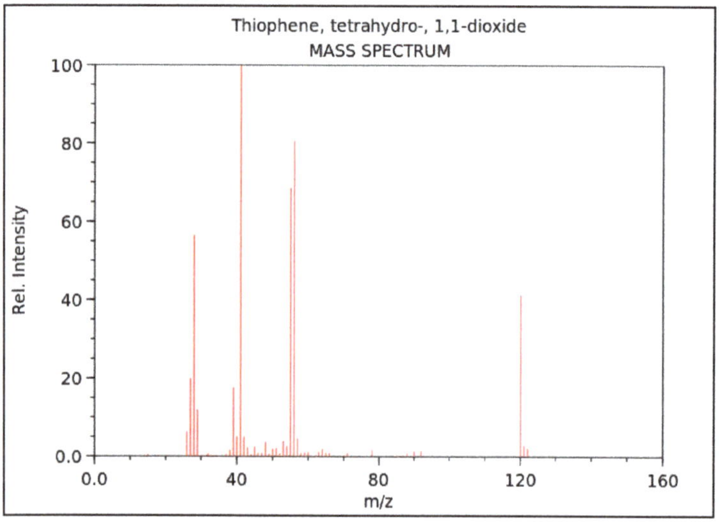

Figure 2. Full-scan mass spectrum of sulfolane. The spectrum is taken from the NIST website (Thiophene, Tetrahydro-, 1,1-Dioxide, 2014). The most abundant fragment ions appeared at m/z 41, 56, and 55, and the molecular ion appeared at m/z 120 [68].

In the official method by the government of Alaska, the molecular ion of sulfolane which appears at m/z 120 in the mass spectrum is required to be used as the quantifier ion for sulfolane detection. Additionally, the ions at m/z 41, 55, and 56 are specified as qualifier ions, corresponding to the loss of CH_3SO_2, $C_2H_4SO_2$, and $C_2H_5SO_2$ fragments, respectively [48]. On the other hand, the official method by the government of BC did not specify any particular ion for the detection of sulfolane using MS [47].

4.4. Internal Standard

An internal standard is a compound added at a known level to samples, standards, and blanks to correct for analytical errors. Two approaches are employed for adding internal standards. In the first approach, internal standards are added before sample processing (surrogate standards). This method allows the standards to correct errors at all stages of the analysis, including sample preparation, separation, and final determination. However, it is crucial to choose an internal standard that would not be present in the samples and mimics the partitioning, injection, separation, and detector response profiles of the original compound. This can be challenging when using a non-selective detector, such as FID [17].

In contrast, when using mass spectrometry (MS), isotopically labeled compounds can serve as internal standards. These compounds have the same structure as the analyte, but some of their atoms have been replaced by their stable isotopes. This method provides a close match between the chemical and physical properties of the internal standard and the original compound, but MS can separate it from the analyte due to its different molecular weight [17]. A common isotopically labeled internal standard for sulfolane analysis is sulfolane-d8, which has been employed in several studies listed in Table 2 and is required by the official methods for sulfolane analysis [47,48].

The second approach involves adding the internal standard directly to the final sample before injection and analysis. This type of internal standard addition is mostly used to correct errors introduced by the analytical instrumentation, such as uncertainty in injection volume or differences in analysis conditions at different times. In this case, any compound that would not be present in the samples can serve as an internal standard. Naphtalene-d8 is required as this type of internal standard in the Alaskan government's official method for sulfolane analysis [48].

5. Conclusions

The release of sulfolane into the environment and its subsequent spread through groundwater raise significant health and environmental concerns. To effectively study the fate of sulfolane after release into the environment and mitigate these risks, it is essential to evaluate the extent of contamination at different sites, which necessitates the use of a reliable analytical method. However, many methods in the literature are inefficient or have significant negative environmental impacts. By examining the existing literature on this subject, valuable insights can be gained into the properties of sulfolane and the challenges it may pose in analytical method development. These insights can then be used as a foundation to devise more eco-friendly and efficient methods. Therefore, in this review, various analytical methods for the analysis of sulfolane were thoroughly examined, summarized, and critically assessed.

A notable area where the analytical procedures for sulfolane analysis greatly differ is in the sample preparation step. Not all procedures yielded sufficiently sensitive methods based on regulatory criteria, and many of the acceptable ones required large amounts of chlorinated solvents, which can be hazardous to the environment. However, it might be possible to modify or downscale these procedures to reduce the required solvent volumes for extraction. Moreover, the physical and chemical properties of sulfolane can be exploited to enhance the efficiency of extraction. For example, the Soxhlet extraction method takes advantage of the thermal stability of sulfolane to improve extraction efficiency from soils. While this method demands a significant amount of solvent and a long extraction time, the thermal stability can be leveraged by other high-temperature extraction techniques such as microwave-assisted extraction or pressurized liquid extraction. This could result in enhanced extraction efficiency and speed, and potentially reduced solvent consumption.

The capability of the extraction method to extract sulfolane co-contaminants is another important and often overlooked aspect for evaluating the sample preparation method. Examining co-contaminants alongside sulfolane may provide a better understanding of the origin and fate of the contamination. For instance, both sulfolane and the co-contaminants' migration behavior might be affected by the presence of the other compound. Thus, it is crucial to evaluate the extraction method's capability to extract these compounds simultaneously.

Concerning the final determination of sulfolane, GC-MS and GC-FID were the exclusive techniques employed for separation and analysis in the analytical process. Among the two, GC-MS has proven to be a sufficiently sensitive and selective technique for quantifying sulfolane. Nevertheless, other analytical techniques, such as LC-MS/MS, could be investigated if co-contaminants are not suitable for GC analysis.

In summary, based on the literature reviewed in this study, there is a clear need for a more efficient and environmentally friendly sample preparation procedure for analyzing sulfolane in environmental samples. This method should accommodate the specific requirements of different environmental matrices, including soil or plant samples, and should be adaptable to different analytical instruments. Furthermore, it should be optimized to simultaneously analyze potential co-contaminants.

Author Contributions: Conceptualization, M.K and T.G.; methodology, M.K. and T.G.; writing—original draft preparation, M.K.; writing—review and editing, T.G.; supervision, T.G.; project administration, T.G.; funding acquisition, T.G. All authors have read and agreed to the published version of the manuscript.

Funding: This research was funded by Shell through Morwick G360 Group, University of Guelph.

Data Availability Statement: Not applicable.

Conflicts of Interest: The authors declare no conflict of interest.

References

1. Clark, E. Sulfolane and Sulfones. In *Kirk-Othmer Encyclopedia of Chemical Technology;* John Wiley & Sons, Inc.: Hoboken, NJ, USA, 2000; ISBN 978-0-471-23896-6.
2. Bak, A.; Kozik, V.; Dybal, P.; Kus, S.; Swietlicka, A.; Jampilek, J. Sulfolane: Magic Extractor or Bad Actor? Pilot-Scale Study on Solvent Corrosion Potential. *Sustainability* **2018**, *10*, 3677. [CrossRef]
3. Travis, C.C.; Arms, A.D. Bioconcentration of Organics in Beef, Milk, and Vegetation. *Environ. Sci. Technol.* **1988**, *22*, 271–274. [CrossRef]
4. O'Neil, M.J. (Ed.) *The Merck Index: An Encyclopedia of Chemicals, Drugs, and Biologicals*, 13th ed.; Merck: Whitehouse Station, NJ, USA, 2001; ISBN 978-0-911910-13-1.
5. Coetzee, J.F. Sulpholane: Purification, Tests for Purity and Properties. *Pure Appl. Chem.* **1977**, *49*, 211–216. [CrossRef]
6. Willer, R. Sulfoxides. In *Kirk-Othmer Encyclopedia of Chemical Technology;* Kirk-Othmer, Ed.; Wiley: Hoboken, NJ, USA, 2000; ISBN 978-0-471-48494-3.
7. Greene, E.; Gieg, L.; Coy, D.; Fedorak, P. Sulfolane Biodegradation Potential in Aquifer Sediments at Sour Natural Gas Plant Sites. *Water Res.* **1998**, *32*, 3680–3688. [CrossRef]
8. Tindal, M.; Sevigny, J.; Potter, K.; Roe, S.; Hill, J. *Canadian Environmental Quality Guidelines for Sulfolane: Water and Soil: Scientific Supporting Document;* Canadian Council of Ministers of the Environment: Winnipeg, MA, Canada, 2006; ISBN 978-1-896997-59-9.
9. Meyers, R.A. (Ed.) UOP SULFOLANE PROCESS. In *Handbook of Petroleum Refining Processes;* McGraw-Hill Education: New York, NY, USA, 2016; ISBN 978-0-07-185049-0.
10. Kim, C.; Clarke, W.; Lockington, D. Competitive Adsorption of Sulfolane and Thiolane on Clay Materials. *Korean J. Chem. Eng.* **1999**, *16*, 215–220. [CrossRef]
11. Stewart, O.; Minnear, L. *Sulfolane Technical Assistance and Evaluation Report;* Alaska Department of Environmental Conservation: Anchorage, AK, USA, 2010.
12. Dinh, M.; Hakimabadi, S.G.; Pham, A.L.-T. Treatment of Sulfolane in Groundwater: A Critical Review. *J. Environ. Manag.* **2020**, *263*, 110385. [CrossRef]
13. Khan, M.; Yu, L.; Achari, G. Sulfolane in Contaminated Sites: Environmental Toxicity and Bioremediation Technologies. *Environ. Rev.* **2022**, *30*, 217–227. [CrossRef]
14. Testing Status of Sulfolane 11054. Available online: https://ntp.niehs.nih.gov/go/ts-11054 (accessed on 17 September 2023).
15. Mulder, D.; Yu, L.; Achari, G. A Laboratory and Field Investigation of Aerobic Biodegradation of Sulfolane in Groundwater. *J. Chem. Technol. Biotechnol.* **2021**, *96*, 2865–2871. [CrossRef]
16. Headley, J.V.; Fedorak, P.M.; Dickson, L.C. A Review of Analytical Methods for the Determination of Sulfolane and Alkanolamines in Environmental Studies. *J. AOAC Int.* **2002**, *85*, 154–162. [CrossRef]
17. Skoog, D.A.; West, D.M.; Holler, F.J.; Crouch, S.R. *Fundamentals of Analytical Chemistry*, 10th ed.; Cengage: Boston, MA, USA, 2022; ISBN 978-0-357-45043-7.

18. Groundwater Sampling Operating Procedure. Available online: https://www.epa.gov/sites/default/files/2015-06/documents/Groundwater-Sampling.pdf (accessed on 16 April 2023).
19. Soil Sampling Operating Procedure. Available online: https://www.epa.gov/sites/default/files/2015-06/documents/Soil-Sampling.pdf (accessed on 17 September 2023).
20. Saint-Fort, R. Sulfolane Attenuation by Surface and Subsurface Soil Matrices. *J. Environ. Sci. Health Part A Tox. Hazard. Subst. Environ. Eng.* **2006**, *41*, 1211–1231. [CrossRef]
21. Doucette, W.; Chard, J.; Moore, B.; Staudt, W.; Headley, J. Uptake of Sulfolane and Diisopropanolamine (DIPA) by Cattails (Typha Latifolia). *Michrochem. J.* **2005**, *81*, 41–49. [CrossRef]
22. Lee, J.K.; Lee, W.J.; Cho, Y.-J.; Park, D.H.; Lee, Y.-W.; Chung, J. Variation of Bacterial Community Immobilized in Polyethylene Glycol Carrier during Mineralization of Xenobiotics Analyzed by TGGE Technique. *Korean J. Chem. Eng.* **2010**, *27*, 1816–1821. [CrossRef]
23. Izadifard, M.; Achari, G.; Langford, C. Degradation of Sulfolane Using Activated Persulfate with UV and UV-Ozone. *Water Res.* **2017**, *125*, 325–331. [CrossRef] [PubMed]
24. Dominic, J.A.; Somathilake, P.; Achari, G.; Langford, C.H.; Tay, J.-H. Sunlight Mediated Passive Wastewater Treatment Technology Using Photochemical Reduction of Ferric Iron for Decontamination of Various Aqueous Contaminants. *Sol. Energy* **2018**, *173*, 470–477. [CrossRef]
25. Izadifard, M.; Achari, G.; Langford, C. Mineralization of Sulfolane in Aqueous Solutions by Ozone/CaO2 and Ozone/CaO with Potential for Field Application. *Chemosphere* **2018**, *197*, 535–540. [CrossRef]
26. Heydari, G.; Langford, C.; Achari, G. Passive Solar Photocatalytic Treatment of Emerging Contaminants in Water: A Field Study. *Catalysts* **2019**, *9*, 1045. [CrossRef]
27. Khan, M.; Yu, L.; Achari, G.; Tay, J. Degradation of Sulfolane in Aqueous Media by Integrating Activated Sludge and Advanced Oxidation Process. *Chemosphere* **2019**, *222*, 1–8. [CrossRef]
28. Khan, M.; Yu, L.; Tay, J.; Achari, G. Coaggregation of Bacterial Communities in Aerobic Granulation and Its Application on the Biodegradation of Sulfolane. *J. Hazard. Mater.* **2019**, *377*, 206–214. [CrossRef]
29. Khan, M.; Yu, L.; Achari, G. Field Evaluation of a Pressurized Ozone Treatment System to Degrade Sulfolane in Contaminated Groundwaters. *J. Environ. Chem. Eng.* **2020**, *8*, 104037. [CrossRef]
30. Khan, M.; Yu, L.; Hollman, J.; Tay, J.; Achari, G. Integration of Aerobic Granulation and UV/H2O2 Processes in a Continuous Flow System for the Degradation of Sulfolane in Contaminated Water. *Environ. Sci. Water Res. Technol.* **2020**, *6*, 1711–1722. [CrossRef]
31. Yang, Y.; Yu, L.; Gopal, A. Aerobic Biodegradation of Sulfolane Using Archaea and Pseudomonas Strains. *J. Chem. Technol. Biotechnol.* **2022**, *97*, 1763–1770. [CrossRef]
32. Yu, L.; Iranmanesh, S.; Keir, I.; Achari, G. A Field Pilot Study on Treating Groundwater Contaminated with Sulfolane Using UV/H2O2. *Water* **2020**, *12*, 1200. [CrossRef]
33. Dharwadkar, S.; Yu, L.; Achari, G. Enhancement of LED Based Photocatalytic Degradation of Sulfolane by Integration with Oxidants and Nanomaterials. *Chemosphere* **2021**, *263*, 124. [CrossRef] [PubMed]
34. Dharwadkar, S.; Yu, L.; Achari, G. Photocatalytic Degradation of Sulfolane Using a LED-Based Photocatalytic Treatment System. *Catalysts* **2021**, *11*, 624. [CrossRef]
35. Yu, L.; Mehrabani-Zeinabad, M.; Achari, G.; Langford, C. Application of UV Based Advanced Oxidation to Treat Sulfolane in an Aqueous Medium. *Chemosphere* **2016**, *160*, 155–161. [CrossRef] [PubMed]
36. Kasanke, C.; Leigh, M. Factors Limiting Sulfolane Biodegradation in Contaminated Subarctic Aquifer Substrate. *PLoS ONE* **2017**, *12*, e0181462. [CrossRef] [PubMed]
37. Chang, S.; Wu, C.; Yang, C.; Lin, C. Evaluation Use of Bioaugmentation and Biostimulation to Improve Degradation of Sulfolane in Artificial Groundwater. *Chemosphere* **2021**, *263*, 127919. [CrossRef] [PubMed]
38. Stevens, S.; Richardson, D. Analysis of Sulfolane in Biochar. *Appita* **2018**, *71*, 349–353.
39. Eckardt, M.; Greb, A.; Simat, T.J. Polyphenylsulfone (PPSU) for Baby Bottles: A Comprehensive Assessment on Polymer-Related Non-Intentionally Added Substances (NIAS). *Food Addit. Contam. Part Chem. Anal. Control Expo. Risk Assess.* **2018**, *35*, 1421–1437. [CrossRef]
40. Headley, J.V.; Peru, K.M.; Dickson, L.C. Gas Chromatographic–Mass Spectrometric Determination of Sulfolane in Wetland Vegetation Exposed to Sour Gas-Contaminated Groundwater. *J. Chromatogr. A* **1999**, *859*, 69–75. [CrossRef]
41. Headley, J.; Dickson, L.; Peru, K. Comparison of Levels of Sulfolane and Diisopropanolamine in Natural Wetland Vegetation Exposed to Gas-Condensate Contaminated Ground Water. *Commun. Soil Sci. Plant Anal.* **2002**, *33*, 3531–3544. [CrossRef]
42. Versace, F.; Uppugunduri, C.R.S.; Krajinovic, M.; Théorêt, Y.; Gumy-Pause, F.; Mangin, P.; Staub, C.; Ansari, M. A Novel Method for Quantification of Sulfolane (a Metabolite of Busulfan) in Plasma by Gas Chromatography–Tandem Mass Spectrometry. *Anal. Bioanal. Chem.* **2012**, *404*, 1831–1838. [CrossRef] [PubMed]
43. Silinski, M.A.R.; Uenoyama, T.; Cooper, S.D.; Fernando, R.A.; Robinson, V.G.; Waidyanatha, S. Development and Validation of an Analytical Method for Quantitation of Sulfolane in Rat and Mouse Plasma by GC-MS. *J. Anal. Toxicol.* **2019**, *43*, 477–481. [CrossRef] [PubMed]
44. Shipkowski, K.A.; Cora, M.C.; Cesta, M.F.; Robinson, V.G.; Waidyanatha, S.; Witt, K.L.; Vallant, M.K.; Fallacara, D.M.; Hejtmancik, M.R.; Masten, S.A.; et al. Comparison of Sulfolane Effects in Sprague Dawley Rats, B6C3F1/N Mice, and Hartley Guinea Pigs after 28 Days of Exposure via Oral Gavage. *Toxicol. Rep.* **2021**, *8*, 581–591. [CrossRef] [PubMed]

45. Greene, E.; Fedorak, P. Nutrient Stimulation of Sulfolane Biodegradation in a Contaminated Soil from a Sour Natural Gas Plant and in a Pristine Soil. *Environ. Technol.* **2001**, *22*, 619–629. [CrossRef] [PubMed]
46. Brandao, M.; Yu, L.; Garcia, C.; Achari, G. Advanced Oxidation Based Treatment of Soil Wash Water Contaminated with Sulfolane. *Water* **2019**, *11*, 2152. [CrossRef]
47. BC Lab Manual for Sulfolane Analysis 2017, Government of British Columbia (Canada). Available online: https://www2.gov.bc.ca/assets/gov/environment/research-monitoring-and-reporting/monitoring/emre/methods/sept2017/bc_lab_manual_sulfolane_method_15sept2017.pdf (accessed on 26 December 2023).
48. Sulfolane Key Elements Document, Version 4. Available online: https://dec.alaska.gov/media/6165/key-element-all-media-2013-7-22.pdf (accessed on 26 December 2023).
49. Fedorak, P.; Coy, D. Biodegradation of Sulfolane in Soil and Groundwater Samples from a Sour Gas Plant Site. *Environ. Technol.* **1996**, *17*, 1093–1102. [CrossRef]
50. Luther, S.; Dudas, M.; Fedorak, P. Sorption of Sulfolane and Diisopropanolamine by Soils, Clays and Aquifer Materials. *J. Contam. Hydrol.* **1998**, *32*, 159–176. [CrossRef]
51. Greene, E.A.; Coy, D.L.; Fedorak, P.M. Laboratory Evaluations of Factors Affecting Biodegradation of Sulfolane and Diisopropanolamine. *Bioremediation J.* **1999**, *3*, 299–313. [CrossRef]
52. Kim, C.; Clarke, W.; Lockington, D. Determination of Retardation Coefficients of Sulfolane and Thiolane on Soils by K-Ow-K-Oc and Solubility Parameter, Batch and Column Experiments. *Environ. Geol.* **2000**, *39*, 741–749. [CrossRef]
53. Kim, C.; Lockington, D.; Clarke, W.; Kim, C. Preliminary Determination of Pollutants Plume in Groundwater at Hazardous Solid Waste Disposal Site by Employing CPT and Rig. *Environ. Technol.* **2000**, *21*, 17–30. [CrossRef]
54. Kim, C.; Chon, B. Oxygen Uptake Characteristics of Soil Inoculum Amended with Thiophene Derivatives. *Korean J. Chem. Eng.* **2002**, *19*, 773–779. [CrossRef]
55. Yang, C.; Liu, S.; Su, Y.; Chen, Y.; Lin, C.; Lin, K. Bioremediation Capability Evaluation of Benzene and Sulfolane Contaminated Groundwater: Determination of Bioremediation Parameters. *Sci. Total Environ.* **2019**, *648*, 811–818. [CrossRef] [PubMed]
56. Lin, C.-W.; Liu, S.-H.; Wu, C.-F.; Chang, S.-H. Critical Factors for Enhancing the Bioremediation of a Toxic Pollutant at High Concentrations in Groundwater: Toxicity Evaluation, Degrader Tolerance, and Microbial Community. *J. Environ. Manag.* **2021**, *277*, 111487. [CrossRef] [PubMed]
57. Li, Z.; Song, J.; Zhang, S.; Wang, J.; Jian, X. Measurement and Correlation of Liquid-Liquid Equilibrium for the Ternary System (Water+1,2-Dichloroethane plus Sulfolane) at 288.15, 298.15, and 308.15 K. *Chin. J. Chem. Eng.* **2022**, *51*, 109–114. [CrossRef]
58. Tobiszewski, M.; Namieśnik, J.; Pena-Pereira, F. Environmental Risk-Based Ranking of Solvents Using the Combination of a Multimedia Model and Multi-Criteria Decision Analysis. *Green Chem.* **2017**, *19*, 1034–1042. [CrossRef]
59. Hossaini, R.; Chipperfield, M.P.; Montzka, S.A.; Leeson, A.A.; Dhomse, S.S.; Pyle, J.A. The Increasing Threat to Stratospheric Ozone from Dichloromethane. *Nat. Commun.* **2017**, *8*, 15962. [CrossRef]
60. Pawliszyn, J.; Lord, H.L. (Eds.) *Handbook of Sample Preparation*; Wiley: Hoboken, NJ, USA, 2010; ISBN 978-0-470-09934-6.
61. EPA Method 3550C. Available online: https://www.epa.gov/sites/default/files/2015-12/documents/3550c.pdf (accessed on 17 April 2023).
62. EPA Method 3540C. Available online: https://www.epa.gov/hw-sw846/sw-846-test-method-3540c-soxhlet-extraction (accessed on 26 December 2023).
63. Neely, B.J.; Wagner, J.; Robinson, R.L., Jr.; Gasem, K.A.M. Mutual Solubility Measurements of Hydrocarbon–Water Systems Containing Benzene, Toluene, and 3-Methylpentane. *J. Chem. Eng. Data* **2008**, *53*, 165–174. [CrossRef]
64. Altshuller, A.P.; Everson, H.E. The Solubility of Ethyl Acetate in Aqueous Electrolyte Solutions. *J. Am. Chem. Soc.* **1953**, *75*, 4823–4827. [CrossRef]
65. Yang, Y.; Yu, L.; Iranmanesh, S.; Keir, I.; Achari, G. Laboratory and Field Investigation of Sulfolane Removal from Water Using Activated Carbon. *J. Environ. Eng.* **2020**, *146*, 1680. [CrossRef]
66. Skoog, D.A.; Holler, F.J.; Crouch, S.R. *Principles of Instrumental Analysis*, 7th ed.; Cengage Learning: Melbourne, Australia, 2018; ISBN 978-1-305-57721-3.
67. Kovacevik, B.; Zdravkovski, Z.; Mitrev, S. Pesticide analysis in water samples using GC-MS pulsed splitless injection. *Comptes Rendus L'Académie Bulg. Sci.* **2016**, *69*, 815–820.
68. Thiophene, Tetrahydro-, 1,1-Dioxide. Available online: https://webbook.nist.gov/cgi/inchi?ID=C126330&Mask=200 (accessed on 22 September 2023).

MDPI AG
Grosspeteranlage 5
4052 Basel
Switzerland
Tel.: +41 61 683 77 34

Separations Editorial Office
E-mail: separations@mdpi.com
www.mdpi.com/journal/separations

www.ingramcontent.com/pod-product-compliance
Lightning Source LLC
LaVergne TN
LVHW070447061126
851858LV00011B/176
9783725885077